LIGHTNING
OFTEN
STRIKES
TWICE

Also by Brian Clegg and published by Michael O'Mara Books

10 Short Lessons in Time Travel (Pocket Einstein Series)

LIGHTNING OFTEN

STR KES
TW CE

THE 50 BIGGEST
MISCONCEPTIONS IN
SCIENCE

BRIAN CLEGG

Michael O'Mara Books Limited

First published in Great Britain in 2022
by Michael O'Mara Books Limited
9 Lion Yard
Tremadoc Road
London SW4 7NQ

A CIP catalogue record for this book is available from the British Library.

Papers used by Michael O'Mara Books Limited are natural, recyclable products made from wood grown in sustainable forests. The manufacturing processes conform to the environmental regulations of the country of origin.

ISBN: 978-1-78929-425-5 in hardback print format
ISBN: 978-1-78929-426-2 in ebook format

1 2 3 4 5 6 7 8 9 10

www.mombooks.com

Designed and typeset by Claire Cater
Illustrations by Peter Liddiard

Printed and bound by CPI Group (UK) Ltd, Croydon, CR0 4YY

CONTENTS

Contents

INTRODUCTION

For millennia, folklore and proverbial wisdom have been used in an attempt to explain the world around us. Some of these beliefs, derived from experience, have later proved to have a basis in science – for example, red sky at night really does suggest good weather next morning. Similarly, willow bark does indeed reduce pain, as it contains salicin, which in the form of salicylic acid is better known as aspirin.

Other explanations and beliefs are fictions, and have managed to linger through to the present despite plenty of scientific evidence to the contrary. And sometimes this proverbial wisdom has versions that are directly contradictory. Think, for example, of 'Many hands make light work' and 'Too many cooks spoil the broth'. Or it may have become a ritual that becomes detached from any sense of reality. A great example of this is Groundhog Day in America (the event, rather than the film). Taking place on 2 February every year, the legend has it that the actions of a groundhog (the best known being the semi-mythical Punxsutawney Phil) coming out of its burrow on that day predict what the weather will be like for the following six weeks. If it is overcast and the groundhog casts no shadow, then things are set fine. But if it's sunny, the groundhog is supposed to be scared back into its burrow by the sight of its own shadow, bringing on six weeks of wintry weather.

In this book, we explore fifty commonly held beliefs about our world that are either misleading or downright false. Some are dependent on folklore – take, for instance, the suggestion alluded to in the title of this book that lightning never strikes the same place twice, once so widely believed that it became a way to refer to an event that is unlikely to recur. Others, often featuring in more modern folklore, come from what feels like a more scientific source.

When first attempts at what we would now call science were being developed in ancient times, the standards for producing an explanation or theory were far less rigorous than is the case now. Ancient philosophers had a tendency to make 'scientific' statements based primarily on argument, rather than on making a detailed examination of nature. The great ancient Greek philosopher Aristotle, for example, infamously asserted that women had fewer teeth than men. Simply counting teeth would have shown that this was a false premise – but Aristotle's authority as a philosopher (which held for many centuries) meant that many accepted the assertion as fact. As a result of long acceptance, while not all the ideas from the pre-scientific era were incorrect, we do get ancient errors recurring even today. With the slight proviso that taste and touch could be related, it was Aristotle who confidently asserted that there are five senses (sight, hearing, taste, smell and touch). We are still taught this at school despite it being long disproved.

Another possibility for misunderstanding comes in the form of modern myths that are spread by popular culture.

Take, for example, the belief that eating sugary foods makes children hyperactive. This is strongly reflected in TV shows, from episodes of *The Simpsons* (where somehow European chocolate is considered far more intense in its impact) to *Modern Family*. It feels sensible that 'energy-giving' sugar would make children get overexcited – and it is often presented as if it were a scientific fact. Yet studies have been undertaken that conclusively show this is not the case. Once a pseudoscientific belief becomes part of the culture, however, it can be difficult to shake.

Often, such incorrect beliefs do no real harm, other than spreading misinformation. As it happens, sugar isn't great for children, so even though the reasoning for keeping sugar consumption down to avoid hyperactive behaviour isn't valid, the belief does no damage. In other cases, though, such beliefs can prove distinctly dangerous. The fifty examples here are not in this category. The aim is to entertain and inform. But there have been dangerous beliefs: from the early idea that smoking tobacco was good for you to more recent suggestions that MMR vaccinations cause autism, which had a devastating impact on the lives of some of those who believed them – in the case of the MMR vaccinations, facilitating the spread of measles among children who weren't inoculated, a disease that can cause brain damage and death.

Lightning Often Strikes Twice, by contrast, presents us with misleading concepts where the reality provides surprise and delight. Each topic offers a fascinating opportunity to find out more about science and to challenge what has been assumed to be true. The aim of this book is to bring the stories behind these misunderstandings and myths to life with a clear picture of the realities that lie behind them.

— 1 —

LIGHTNING NEVER STRIKES TWICE

Lightning is a terrifying force of nature. There was a time when these dramatic flashes and bangs in the sky (thunder is just the sound made by lightning as it rips through the air, not a separate phenomenon) were considered to be the action of gods. But although we now know that lightning is caused by a build-up of electrical charge in clouds, probably due to ice particles bashing into each other, rubbing charged electrons away from atoms, it certainly is a phenomenal source of energy.

A typical lightning flash carries a similar amount of energy to the output of a mid-sized power station over the period of a second – but that energy is emitted far quicker. As the electrical energy is discharged it sends air molecules flying at such speeds that localized air temperatures can reach between 20,000 and 30,000 °C (36,000 and 54,000 °F) – over four times the temperature of the surface of the sun. It's this blast, ripping through the air molecules, that forms the distinctive rumble and crash of thunder.

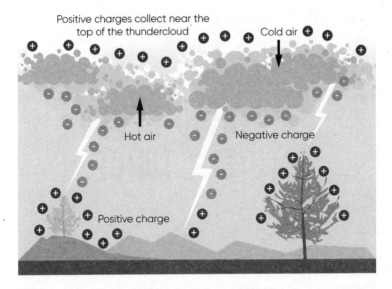

Although in any particular location we might not see thunderstorms too often, they are not at all uncommon. As you read this, there are probably around two thousand on the go around the world, with an average of 8 million lightning strikes taking place each day. (There tend to be more in summer – but it's always summer somewhere.) Most bolts of lightning travel from cloud to cloud and never reach the ground, but it's the strikes that link the clouds to the Earth that give lightning its fearsome reputation and most devastating outcomes – blasting trees, starting fires and killing humans and animals.

With the awareness of the dangers arising from a lightning strike came attempts to reduce the risk. Now, we might expect a lightning conductor, also known as a lightning rod, to be used on a tall building. Ever since the days of Benjamin Franklin there have been two theories for how these work. Such rods may lessen the chances of a strike ever happening by reducing the difference in voltage between the sky and the rooftop when a voltage is induced in the rod, or they may lead the lightning discharge away to the Earth down a safe path. In practice, there is limited evidence that either mechanism really works. Before their development in the eighteenth century, though, there was another, even more dubious option, based on the 'lightning never strikes twice' premise, known as a thunderstone.

This medieval preventative measure involved using a stone that was thought to have already been struck by lightning. This would be placed in a location of risk – for example, up the chimney of a house, where a lightning strike would have a high possibility of setting a thatched roof alight. These stones were most often in reality Stone Age axe heads, but the shape of the stone was assumed to be the result of a lightning strike. Put a stone in place and lightning's aversion to returning to the same location would provide protection.

More often than not, 'Lightning never strikes twice' is used not about lightning itself, but rather proverbially as a way of suggesting that something is unlikely to happen again. Although the first use of the saying hasn't been pinned down for sure, it seems to date back to the nineteenth century. It appears, for example, in an 1851 Australian newspaper and more graphically in the 1860 US novel *Thrilling Adventures of the Prisoner of the Border* by P. Hamilton Myers. Here, the protagonists have just survived a near miss from a cannonball. One says to the other: 'Never fear, Brom. Sit down on it, if you wish to be safe. Lightning never strikes twice in the same place, nor cannon balls either, I presume.'

It's actually obvious that the 'doesn't strike twice' myth can't have any basis in fact. How could a random electrical current possibly know where lightning has struck before? Short of having Zeus or Thor keeping track of their targets, it's not a credible defence against a strike.

In reality, susceptible locations do get struck with remarkable regularity – the Empire State Building, for example, has had as many as fifteen strikes in a single storm and is regularly hit around twenty-five times a year. The failure of this theory even applies to people. US park ranger Roy Sullivan entered the *Guinness Book of Records* as the person who has been hit most often by a lightning bolt – a total of seven times. And he survived every one of them.

– 2 –
WE HAVE
FIVE SENSES

As I mentioned in the introduction to this book, we still teach children at school today that human beings have five senses: sight, hearing, smell, taste and touch. In reality, it's not entirely clear how many senses we do have, as some of the distinctions between them can be hard to pin down, but the total number is certainly more than five.

The familiar senses above were first identified in ancient times. The ancient Greek philosopher Aristotle gave us the famous five, though he was not sure about whether to separate taste and touch, both of which require contact. (Having either four or five fitted well with his theory of the elements – Aristotle concurred with the four earthly elements being earth, water, air and fire, adding in a fifth heavenly element, sometimes called the quintessence.) Aristotle based his theories on experience and argument. It's true that his five senses are the most obvious ones, but it's hard to imagine how he managed to miss one other.

If you hold your hand close to a hot object that is not glowing with heat – the base of an iron, for example – you can tell that it is hot without touching it. That's just as well, as touching something hot causes damage. It's a useful natural protection. But which sense are you using to detect that radiant heat? It is clearly not sight, as something becomes detectably hot long before its temperature is high enough for it to glow visibly. You can't hear, or smell[1] or taste the heat. You are able to detect the heat because of your sixth sense: thermoreception.

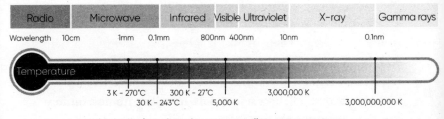

Radio	Microwave		Infrared	Visible	Ultraviolet	X-ray	Gamma rays

Wavelength 10cm   1mm   0.1mm   800nm 400nm   10nm   0.1nm

Temperature

3 K - 270°C   300 K - 27°C   3,000,000 K
30 K - 243°C   5,000 K   3,000,000,000 K

Light 'colours' and corresponding temperatures.

To see how this works, we need to take a step back and consider what heat is. Radiant heat – the kind we are talking about here – is a form of light. We are used to light being visible, but the light that we can see is just a small section in the middle of the whole electromagnetic spectrum, which runs from low-energy radio waves all the way up to X-rays and gamma rays. Light photons that are a little too low in energy for our eyes to detect

---

1.   It *is* sometimes possible to smell that an object is hot, but this is because the heat is causing substances on the object's surface to burn or evaporate. You can't smell the heat itself.

are known as infrared. But though we can't see infrared, our skin can detect it. The detection is crude – it is very localized and lacks clear focus – but a distinctive different sense is in action. This is down to special neurons containing thermoreceptors – you have them in your skin for detection of both heat and cold.

Let's imagine another circumstance that shows the presence of another sense. You are on a theme park ride being spun and dropped and generally accelerated and decelerated. How do you know this is happening if you have your eyes closed? The sense of touch is certainly involved – typically the motion will push you into different parts of your seat or the restraint. But even without this, your body knows that it is being accelerated. A fluid accelerometer inside your head keeps track of what is happening to help you keep your balance. This is not the job of any of the traditional five senses.

Another example is a sense known as proprioception. You can test this out right now. Close your eyes, then touch your nose. Most people can do so easily – but what sense did you use to find where your nose is? Clearly it could not have been any of the traditional five. Proprioception is an awareness of the location of parts of the body that is essential for managing our interaction with the world around us.

And what about pain? In some cases, this seems to be an extension of touch. Touch enables us to detect pressure on our skin and if that pressure becomes too intense, the sensation transitions into pain. But what about, say, the pain of a headache? This clearly isn't a response to a touch, but a totally different kind of sensory stimulus from the triggering of nerves.

We have a collection of other, more subtle, sensory

abilities – common estimates for the total number of senses range from the low twenties to around thirty-three, while the psychologist Michael J. Cohen puts it as high as fifty-three. To do this, Cohen has to resort to what many might consider cheating – for example, considering the sense of air on the skin as being different from the sense of touch. But some animals are able to go far beyond us. Sharks can sense electrical fields from living things, while pigeons use the Earth's magnetic field in their navigation. And even though a bat's echolocation does make use of sound, its sonar-like capabilities are a totally distinct sensory mechanism from its conventional hearing, producing an ability that is closer to sight than to hearing.

– 3 –

THE NORTH STAR IS THE BRIGHTEST STAR IN THE NIGHT SKY

The North Star – properly known as Polaris – has a special place in the night sky of the Northern Hemisphere. The motion of the stars in the heavens appears to be centred on Polaris ('appears to be' in that the motion we see is caused by the rotation of the Earth). The North Star provides a valuable directional guide to those lacking a compass. Because of its practical and astronomical importance, Polaris has gained the reputation of being particularly bright. In reality, though, it isn't even in the top ten of the stars we can see, coming in at number forty-nine if we include our nearest stellar neighbour, the sun.

We need to be a little careful about what is meant by the term 'brightest star'. Historically, the night-time stars included the planets, the so-called 'wandering stars' – but the latter are relatively small bodies in our solar system, illuminated by the light of the sun. The sun, by contrast, is itself a star, close enough to show us just how energetic it is. It is also big by the standards of everything else around it – over 99 per cent of the stuff of the solar system is found in the sun. A star, then, is a vast astronomical body that glows with light, generated by the nuclear reactions that power it.

It might seem there is no need to define 'brightest' – but the problem is that not all stars are the same distance away. The further a luminous object is from us, the fainter it looks. The brightness drops off with the square of its distance away. So, each star has not one, but two different values for its brightness (or magnitudes, the term that astronomers prefer). There is its apparent brightness, and its actual or absolute brightness. The apparent brightness is how intense the star looks to be in the night sky, while the absolute brightness is how bright it would be compared to other stars if they were all the same distance away. (The standard distance astronomers use for this comparison is an arbitrary one of 10 parsecs – about 32.6 light years, which is about 192,000,000,000,000 miles or 300,000,000,000,000 kilometres.)

If we set aside the moon and the planets as not being stars in the modern sense, the brightest star in the night sky is Sirius, the so-called Dog Star, called this because it stands out in the constellation known as Canis Major (the greater dog[2]). One of the reasons that Sirius is so bright is that it is relatively close to us: only around 8.6 light years away. Polaris, by contrast, is considerably more distant at 433 light years.

---

2. For some reason, 'major' and 'minor' in the Latin names of constellations tend to be translated as great and little – but these in Latin would be *magna* and *parva*. The words mean greater and lesser or smaller.

In absolute terms it is a hundred times brighter than Sirius, but it appears fainter because of that extra distance. What we see as the single star Polaris is actually three stars in the same system, but one dominates the others. This yellow supergiant is a variable star – its brightness varies with a period of about four days. Polaris is located in the constellation of Ursa Minor, the lesser bear.

The familiar shape of the Big Dipper (also known as the Plough), part of Ursa Major, points to Polaris in the Little Dipper or Ursa Minor.

Polaris is currently the closest bright star to the 'celestial pole' – the position in the night sky around which the other stars appear to move – but this will not remain the case for ever. As we have seen, the motion of the stars through the sky is caused by the rotation of the Earth – and the axis around which the

Earth rotates is gradually changing direction. The direction in which the axis points moves around a circular path over a period of around 26,000 years (a phenomenon that used to be known poetically as the precession of the equinoxes). During that period, around fourteen different stars will take Polaris's place as the North Star, but it will remain the best pointer to the north for another thousand years yet.

— 4 —

FINGERTIPS WRINKLE IN THE BATH WHEN THEY ABSORB WATER

We all know how our fingertips (and toes) become soft, taking on wrinkly textures, when we're in the bath. Many people assume that this is because the skin has become swollen as a result of absorbing water. But our hands and feet are just as waterproof as the rest of our skin. We've known for a long time that skin keeps water out, but exactly how it worked wasn't discovered until 2012. It's down to fatty molecules called lipids – the best-known lipid is cholesterol, which is one of the types in the skin, along with fatty acids and ceramides.

The structure of a lipid molecule is a long chain, with a head that attracts water and two tails that repel water, due to their relative electrical charges. Often the lipid is bent over into a hairpin shape with the two tails heading in the same direction – but in the protective layer of the skin, the water-repelling tails point in opposite directions, pushing water molecules away however they approach.

So why does the skin on hands and feet react so differently to water to the rest of the body? (Thankfully – it really wouldn't be pleasant if we went wrinkly like that all over.) The presence of water on the skin triggers a nervous system reflex. It is nothing to do with water somehow getting beneath the surface of the skin. In fact, if there is nerve damage, the wrinkling doesn't occur, showing it is an active response of the body.

The most likely reason for this effect is that in wet conditions, things get slippery. When wet, our hands and feet undergo this nervous system reaction that enables them to get a better grip. A good parallel is the design of car tyres. On a totally dry road surface, it's best if a tyre has no indentations like those used in Formula One races in the dry. That way, the maximum amount of rubber is in contact with the road, giving the best possible grip. Yet ordinary car tyres have 'tread' – indentations in the surface that reduce the amount of rubber touching the road.

The reason this reduction is desirable is that the channels in the surface of the tyre allow water to escape, pushing it out, reducing the amount of slippery water between the tyre and the road. This results in much better grip in dangerous wet conditions. (It is also why a traditional tyre struggles with ice, as it has reduced surface but no advantage from pushing aside the slippery stuff.) Much in the same way that the tread on a tyre prevents skidding in the rain, so our wrinkly fingers and toes appear to help us to avoid slips and dropping things.

This explanation was tested out in experiments in which people were asked to pick up a range of objects where both the object and the hands could either be wet or dry. A typical object might be a glass marble. The wrinkly wet fingers gave a better grip on slippery wet marbles, but there was no advantage when trying to pick up dry marbles. There has been some doubt cast on this theory from another experiment in which the results weren't fully reproduced, but it still remains the best-supported current theory.

It seems, then, that those wrinkles on the fingers help us pick things up in the wet, while the wrinkly feet act like tyre treads to maximize our chances of staying upright in wet conditions – provided we walk without shoes and socks on.

— 5 —

WATER IS A GOOD CONDUCTOR OF ELECTRICITY

Bond fans may remember Sean Connery's opening scene in the film *Goldfinger*, where the MI6 operative drops an electric heater into a bath to kill a would-be assassin. We all know that mixing electricity and bath water can have deadly consequences. Yet, remarkably, pure water is a bad conductor of electricity.

Some of our terminology for describing electricity, dating back to the early days of its being harnessed, sounds like we're dealing with water. We speak, for example, of an electrical current. There are clearly differences between electrical systems and plumbing – we don't (thankfully) find electricity draining out of our sockets when there is no plug in them. However, an electric current really does involve something flowing through a wire or other substance: electrically charged particles.

When electricity passes through a wire, those charged particles are electrons. These infinitesimally small particles are usually found in a cloud of activity around the outside of atoms. But in an electrical conductor like a metal, some of the electrons are so loosely attached that when there is a 'potential difference' across the conductor – a voltage – the electrons can flow through the lattice structure of the metal, producing an electrical current.

If a substance has no such loose charged particles available, then it is an insulator – it will not conduct unless the voltage put across it is so huge that it can rip electrons off the otherwise stable atoms. So, for example, air is an insulator. But if the voltage across an air gap at atmospheric pressure exceeds around 30,000 volts per centimetre (76,000 volts per inch), some electrons are pulled loose, and an electrical spark can cross the gap.

Water is a better insulator than air. It takes a huge potential difference of around 700,000 volts per centimetre (1.78 million volts per inch) to cause the resistance to break down. And yet it certainly isn't a good idea to introduce electricity into water, James Bond style. What's more, we know that various animals make use of electrical currents in water. Sharks, for example, can detect life from the tiny voltages generated in living bodies by the electrochemical processes that keep us alive. And some fish can even generate a zap of electricity, both for signalling and for defence. Electric fish can generate voltages up to around 800 volts – but that's nowhere near enough to break down water and make it conduct.

Why, then, is water an insulator, yet we experience water conducting? Because of our confusing use of the term 'water' to include not only pure H_2O, but water plus impurities. Seawater, for example, is obviously not pure water – but even fresh water has chemicals dissolved in it to a lesser extent, notably salts such as chlorides and fluorides. It is only specially treated water, such as distilled water, that is pure enough to be an insulator.

If we take seawater as the most obvious example, most of us, if asked, would say that this salty liquid contains common salt – sodium chloride. But the secret to its electrical conductivity is that seawater doesn't contain sodium chloride at all. When seawater is evaporated in the enormous pans used for sea salt production it *produces* sodium chloride – but the chemical wasn't there initially.

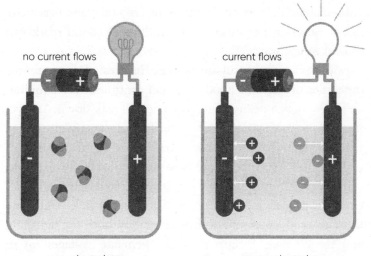

no current flows

current flows

non-electrolyte
(pure water)

strong electrolyte
(sodium chloride solution)

Salts like sodium chloride are known as ionic compounds. The chemical is made up not of atoms, but of ions. These are atoms that have gained or lost electrons. Elements that are in the first column of the periodic table – including sodium –

are good at losing an electron, because they have just one on the outside. Atoms where the outside 'shell' of electrons is full – which happens when the loner is lost – are very stable. As a result, sodium likes to lose an electron and become the positively charged sodium ion. Similarly, elements in the penultimate column of the periodic table, such as chlorine, need only to gain one electron to get a full outer shell. As a result, negatively charged chlorine ions form easily.

It's the attraction of positive and negative electrical charge between the ions that holds salt together. But solvents like water are good at prising apart these ionic bonds. When dissolved, sodium chloride separates into positively charged sodium ions and negatively charged chlorine ions, which float around in the water. What we call water is often in fact water plus impurities: it is only because of the electrical charges on these ions that seawater and tap water are good conductors.

— 6 —

HUMANS HAVE EXCEPTIONALLY LARGE BRAINS

In science-fiction movies and comics from the 1950s, aliens and people of the future often have big, bulgy brains, showing how they had evolved beyond the capabilities of contemporary humans. In these stories, Earth is regularly invaded by nasty-looking individuals with brains so enormous that their wrinkly surfaces protrude horribly from alien skulls. But this idea of bulging brains was based on a misunderstanding of the importance of brain size to intelligence.

There is no doubt that humans have mental capabilities that go far beyond those of most animals. Whether it's in terms of intelligence or imagination and creativity, the species *Homo sapiens* is remarkable. Although it's true that many of our capabilities are not unique – a number of other animals, for instance, make limited use of tools – our technology and our ability to modify our environment to enhance survival goes far beyond the capability of any other species and that ability is, without doubt, a function of our very special brains.

It's true, of course, that having a very small brain can limit mental capabilities. Though, say, a fly may seem clever in its ability to avoid being swatted, it is distinctly limited in its thinking capacity. However, when we compare human brain size with that of other animals, while we do pretty well, we are certainly not in the top ranks.

There are a number of ways to measure the capacity of a brain. One is simple weight or volume. *Homo sapiens* does well on this measure compared with other primates such as chimpanzees or gorillas. However, we are outclassed by a range of larger animals. Human brains weigh in at around 1.3 kilograms (3 pounds), but elephants can have 5-kg (11-lb) brains, while sperm whale brains get as big as 8 kg (18 lb). Even when comparing humans, if we were to go on weight alone, we certainly couldn't predict, for example, that Albert Einstein would be a genius.

Not surprisingly, Einstein's brain has been subject to an unusual amount of study. Its history since his death in 1955 is distinctly bizarre. After Einstein's autopsy, his brain was sliced up into over two hundred pieces, each of which was preserved in collodion, a cellulose–derived substance, not unlike a plastic. The brain pieces went missing for over twenty years, eventually discovered stored in alcohol in a pair of cider jars in the pathologist's garage. Although a few small deviations from the average were noted for Einstein's brain, they are potentially dubious, as the scientists who found these deviations knew that the brain was Einstein's and so expected to find something extraordinary. All we can say for sure is that the brain was a little under average weight at 1,230 grams (2.7 lb).

Let's try to be more subtle about what 'big' means here. You could look at the size of the brain relative to the body, but the mammalian winner on that measure is the shrew, not known as an intellectual giant – in fact, small mammals generally score better than big ones on this approach. What's more interesting, though, is just how those brains are made up. A brain is not a homogeneous lump of matter, but consists of a collection of many cells (billions in a human) of different types.

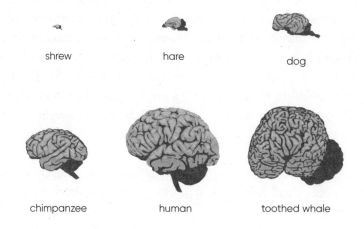

shrew

hare

dog

chimpanzee

human

toothed whale

Relative brain size of different species.

Specifically, the aspect of brain size that seems to most closely reflect intelligence is the number of neurons – the key cells of the brain – in the part of the organ known as the forebrain. Here, humans do relatively well, though we are still beaten significantly by, for example, pilot whales. Men also tend to have more neurons in the forebrain than women, despite there being no indication of a resultant difference in intelligence between sexes.

In reality, it seems likely that the unique aspects of human intelligence reflect a whole mix of factors, rather than having a specific cause. Our brains are certainly larger than we would expect for our body size, with a lot more complexity than might be expected in the forebrain, notably in our having 15 billion neurons in the cerebral cortex and a vast number of connections between them. It seems likely that what makes us the way we are is a combination of brain size, structure and the way that the connections in the brain form. To an extent, though, it remains a mystery.

– 7 –

THERE ARE THREE STATES OF MATTER

That there are three states of matter – solid, liquid and gas – is another 'fact' still taught in schools, despite being fundamentally wrong. To demonstrate just how wrong this description is, if we look at the universe as whole, over 99.9 per cent of all matter is not in any of those three states. That's more than a minor failure. As a statement, 'there are three states of matter' is remarkably inaccurate.

The three states we are taught about are particularly familiar to us when it comes to water – the only substance that is typically present in all three of those states at the temperatures we experience on Earth. In ice – solid water – the molecules that make up the substance are tightly bound together by the force of electromagnetism. They are never totally still – they vibrate – but their positions are relatively fixed in a lattice structure. In liquid water, those strong bonds are broken, but the water molecules are still slow-moving enough to be attracted to each other and to form a fluid that will stay in place under gravity. And in water vapour – gaseous water – the molecules are moving sufficiently fast that the electromagnetic attraction doesn't hold them together and so they can float around, filling an available space. (Note that steam is not just water vapour; it is the droplets of liquid water it contains that makes it visible.)

We are very lucky that water molecules are strongly attracted to each other by so-called hydrogen bonding. The two hydrogen atoms of the H_2O molecule are relatively positively charged, while the single oxygen atom is relatively negatively charged. Because of this, a hydrogen atom from one molecule is attracted to an oxygen atom in another molecule. If it weren't for hydrogen bonding, water would boil at -70 °C (-94 °F). There would be no liquid water on Earth, which means no life.

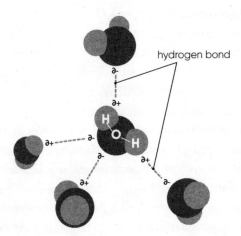

Solid, liquid and gas, then, are the states of matter we are introduced to at school – so what's missing? Primarily, plasma. It is plasma that makes up 99.9 per cent of the universe. This is the main constituent of stars, and the bulk of matter in the universe is found in stars. Plasma is sometimes portrayed as being a special kind of gas – but that's misleading. The difference between gas and plasma is a lot bigger than that between solids, liquids and gases.

All three of the traditional states of matter are composed of atoms (which may be grouped as molecules). The only difference between them is in the way that those atoms interact with each other. But in the transition from gas to plasma, the matter ceases to be made up of atoms – instead, a plasma's component parts are ions. These are atoms that have had electrons added to them or subtracted from them, leaving the ions electrically charged. A plasma can seem like a gas because the ions are free to fly around like atoms are in a gas – but a plasma's behaviour is very different because of those charged particles.

Unlike an ordinary gas, the ions mean that a plasma is good at conducting electricity. This is why plasma is used in some TVs – the plasma screen is made up of a collection of tiny cells containing gas, which is converted to a plasma by a high voltage. Collisions of electrons with particles in the plasma produce ultraviolet photons, which are then used to stimulate coloured phosphors to produce the picture. Other locations where we encounter plasma on the Earth include flames and lightning.

While there is no doubt at all about the four states of matter, most physicists would describe some specialist materials, known as Bose-Einstein condensates, as a fifth state. These are very thin gases of bosons that have been cooled to temperatures very close to absolute zero (-273.15 °C). The 'boson' part refers to a class of particles that includes photons of light, but more importantly here, also nuclei of some atoms. In these condensates, the component particles are at their lowest possible energy, behaving like nothing else. They can be superfluids that flow with no viscosity, and have been demonstrated to be able to slow light to a walking pace, or even to trap it temporarily within the material.

– 8 –

THE EARTH'S POPULATION IS GROWING EXPONENTIALLY
(AND WE'RE DOOMED)

The number of humans on Earth has grown remarkably over the last couple of centuries. In 1800 it was around 1 billion. By 1900 it had doubled. Just sixty years later it was 3 billion. And the next sixty years saw it rise to a remarkable 7.8 billion. This reflects in part the impact that science has had in reducing infant deaths. If the population were to carry on increasing this way, humanity would be in serious trouble.

Back in 1798, English economist Thomas Malthus predicted a disaster on the horizon as populations rose and food production proved unable to keep up with the numbers, leading to mass starvation. Malthus did not envisage population rises of even close to the level we have seen – yet that catastrophe has not happened, again thanks to science and technology.

The vastly improved productivity of agriculture in the last two hundred years has enabled food production to more than keep up with population growth. It is worth reflecting on this when some suggest we need to go back to old ways and purely organic farming. This would result in disastrous food shortages. Of course, there are already regions of the world where food is hard to obtain – but this reflects the difficulties of getting the food to the right places, rather than there being insufficient food overall.

However, even with the help of science and technology, the Earth has a limited capacity – and the rate of growth of the population has been quite staggering. So, is that growth exponential, and, as a result, are we doomed?

Exponential growth is often used to mean 'growing very quickly' – and it often involves explosively fast growth – but that's not what makes it special. A familiar form of growth is linear, where something grows, say, by the same amount every year. So, for example, if the population grew by 1 million people a year, the growth would be linear. We calculate the growth by multiplying the growth by the number of years. To be exponential growth, however, we make the number of years the 'exponent' – the mathematical power a number is raised to.

Take, for example, exponential doubling. If something doubles every year, then the size of it after n years is 2^n times the original value – here n is the exponent. This is often illustrated with a story about a wise man being offered one of those boundless gifts that extravagant (and foolish) kings like to provide in fairy tales. The wise man asks for something that appears to be trivial – some rice. Specifically, he wants to employ exponential doubling, using a chess board to specify how much. He asks for one grain of rice for the first square, two for the second square, four for the third square and so on, until all sixty-four squares are accounted for.

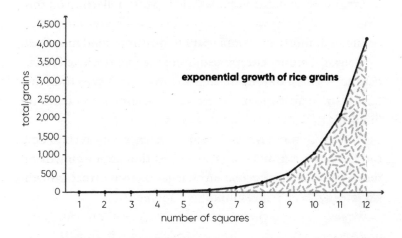

It doesn't sound too bad. But the king has given the go-ahead to award more rice than ever exists at one time, even today, let alone in the medieval period when this story was conceived. The total rice required by the wise man amounts to some 18.5 billion billion grains (around six hundred times current world production). This is because exponential growth rapidly runs away with us.

It is indeed true that the rate of population growth has *been* exponential. If the time it takes for population to double is the same or less than it took in a subsequent period, that's exponential, for example. And that has happened. Population doubled in the hundred years from 1800 to 1900, and more than doubled from 1900 to 2000. But the same isn't true of late. Population is still growing, but the rate of growth is slowing rapidly as smaller families become the norm, in many developed countries now being lower than the average 2.1 children per family to sustain population levels. At the time of writing, the expectation is that population will reach a maximum around the year 2100 of 10 to 12 billion. And then it will fall.

As infant mortality and poverty reduce in a country, so does the average number of children born to a family. In many countries, this number is already below the approximately 2.3 required to sustain the population. Although 10 to 12 billion is a lot of people, it is within the bounds that can be supported by modern agriculture. We are no longer headed for a Malthusian disaster from population explosion. Long term, the problem for humanity is likely to be not having a sufficiently large population.

– 9 –

GOLDFISH HAVE A THREE-SECOND MEMORY

It's a widely accepted 'fact' that goldfish have serious problems remembering things. There's even a joke to go with it: 'They think that I don't mind eating fish flakes over and over because I've only got a three-second memory … Oh, lovely! Fish flakes!'

It's not a great joke – science and hilarity don't always make great bedfellows. The biggest problem with the goldfish joke, though (whether you make the fish's memory last three seconds or opt the equally popular figures of five or nine seconds), is that it bears no resemblance to the truth.

I used to keep goldfish in a pond, feeding them once a day from the same spot. As soon as I turned up at that spot, the fish would start milling around nearby, waiting for the food to arrive. If they forgot everything after just a few seconds, this would not be possible. In 2003, in a classic example of a university study demonstrating something that everyone who owns an animal knows already, researchers at Plymouth University trained fish to get food by pressing a lever, a mechanism that subsequently only worked for an hour a day. This memory was shown to continue for several months.

The year 2003 was clearly a breakthrough one for goldfish studies, as the US–Australian TV show *MythBusters* also undertook experiments with piscine memories, in a programme broadcast in 2004. In their experiments, goldfish remembered colour prompts and routes through a maze for at least a month. And there are apparently papers supporting the effectiveness of fish memory going back as far as 1908.

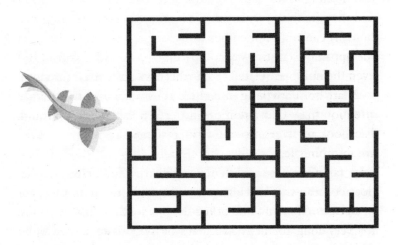

Why, then, is the poor goldfish accused of having such faulty recall? A number of sources suggest that the origin of the myth was a TV advert, but it seems uncertain if this was really the case (and if so, what the advertisement was for or when it was broadcast).

More recently, in 2015, this hoary old myth was revived with a subtly different flavour when Microsoft Canada published a report claiming that as a result of our exposure to digital media, human attention span was dropping. Their paper claimed that in 2000, the average human attention span was 12 seconds. But it was said to have dropped to 8 seconds by 2013 – which the paper tells us is less than the 'average attention span of a goldfish' of 9 seconds. Microsoft's source for this assertion was a website called Statistic Brain. Unfortunately, it hasn't proved possible to find any source beyond that website.

In reality, the assertion is doubly dodgy. First, that old, incorrect goldfish statistic (whether you go for 3, 5 or 9 seconds) had nothing to do with attention span – it was about memory, and they aren't the same thing. No one was ever claiming that humans can only remember things for 8 seconds in 2013, so the apparent statistic was comparing apples and goldfish. But even the numbers related to humans have since been shredded.

Apart from anything else, there is no such thing as average attention span. Our attention span when, say, flicking through Facebook messages, driving a car, reading a book or watching an absorbing feature film are entirely different. We certainly have far more distractions than once was the case. But that doesn't mean that the internet has ruined our attention spans in all cases. Arguably, one of the reasons we don't pay attention for very long to many social media posts is that they don't deserve more than a few seconds of our time.

Next time someone hauls in the memory (or attention span) of the poor old goldfish, you can confidently tell them that they are wrong. At the same time, you can poke fun at the limited memory of the journalist or PR person who has once more launched this myth on the world.

— 10 —

DINOSAURS WENT EXTINCT AFTER AN ASTEROID HIT EARTH

For many millions of years, dinosaurs and their kin were abundantly successful. Human beings have so far existed for about three hundred thousand years. Dinosaurs were around for approximately a hundred and sixty-five million years. It's going to take us a long time to catch up. Yet, as we are all taught at school, dinosaurs suffered an extinction event around 66 million years ago.

The cause of the catastrophe that happened back then was for a long time disputed, but it is now thought to be almost certainly due to the impact on the Earth of an asteroid. Uncovering the evidence for this was the result of some remarkable detective work by the geologist Walter Alvarez and his father, physicist Luis Alvarez.

Walter had been studying the layer in the Earth's crust associated with that great extinction event around 66 million years ago. Because of the way that geological layers build up, the so-called K–Pg boundary (short for Cretaceous–Paleogene, the names of the two periods either side of the extinction event) gives a picture of what was happening to the Earth's surface at this point in the past.

The father and son duo noticed that there was an unusually large amount of the element iridium in the layer. This is a heavy metal, which is usually scarce near the surface, as its density means that it is easily pulled by gravity further into the Earth. Iridium discovered in the Earth's crust, then, mostly comes from meteor impacts. However, Walter and Luis discovered that there was around ninety times as much of the element in the layer as you might expect from the typical arrival rate of extraterrestrial iridium. What's more, this was the case wherever around the world they took samples.

The not-exactly smoking gun for the theory was the discovery of a huge 200-kilometre-wide crater that lies across the edge of the Yucatán Peninsula of Mexico. This vast remnant, known as Chicxulub, has been calculated to be the outcome of a roughly 10-kilometre (6 mile)-wide asteroid hitting the Earth at about 20 kilometres (12 miles) per second. The impact would have been on a scale of 5 billion times the blast from the Hiroshima nuclear bomb. And the result of this impact was terrifying.

The combination of material smashed out of the Earth and falling back, earthquakes, tsunamis and the almost total blanking out of sunlight for several years by the clouds of dust and ash resulted in the destruction of around 75 per cent of animal and plant species. To put this into context, we are currently losing a worrying level of species in what has sometimes been called the sixth mass extinction – but by comparison with the

Chicxulub event, what's happening now is trivial. As yet, we have only lost perhaps 5 per cent of species.

So far then, the statement that the dinosaurs went extinct after an asteroid impact would seem to be accurate. But it's not. Some dinosaurs did survive – and their ancestors, much evolved but still dinosaurs, remain with us today. Perhaps the reason this sounds unlikely is because we have a very particular idea of what dinosaurs look like, thanks to what we see in museums and Jurassic Park movies. And it is true that most of the familiar species, from Tyrannosaurus rex to velociraptors, did not survive. But much of our mental picture of dinosaurs as being smooth or scaly skinned giant lizard-like animals is simply not accurate.

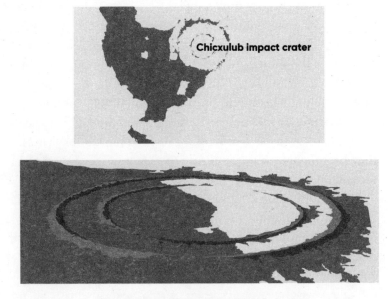

Chicxulub impact crater

The largely disappeared crater from the asteroid impact 66 million years ago.

Unlike lizards, dinosaurs were warm blooded. They laid eggs and many dinosaurs (including those ferocious velociraptors) had feathers. What's more, some dinosaurs could fly. Sound familiar? Although they have, indeed, evolved since the extinction event, birds are dinosaurs. In part because the ancestors of birds were relatively small, they were better able to survive the difficult times that followed the impact than most dinosaur species. Not all dinosaurs went extinct.

— 11 —

PLASTIC WASTE CONTRIBUTES TO CLIMATE CHANGE

Environmental issues are often in the news – and rightly so. It is important that we make the best of our environment for the future of humanity. Often, these issues are framed as 'saving the planet'. This doesn't make any sense. The planet is just fine – and will be so whatever we throw at it. In its 4.6-billion-year history it has gone through periods of extreme cold and extreme heat that are far worse than anything we are likely to experience in the next few thousand years thanks to human involvement. The Earth will bounce back.

However, the same resilience isn't applicable to any particular species, and whether we are thinking of the preservation of *Homo sapiens* or of keeping the biodiversity that both makes our survival more likely and keeps the world the way we like it to be, it is important that we try to ensure that humanity's unique ability to transform the environment around us does not result in catastrophic changes.

The scientific consensus is that we are experiencing climate change, primarily driven by our use of fossil fuels – this is by far the biggest environmental issue we face. As a result of an increase in greenhouse gases in the atmosphere, summers are getting warmer and extreme weather events, from wildfires to flooding, more frequent. Temperature rises also imply increased sea levels. The initial cause for this is simply that water expands and takes up more room as it gets warmer – but we are also seeing increasing melting occurring in long-term ice sheets, putting low-lying countries at risk as sea levels rise.

While there is reasonable dispute over the exact details of how fast changes are happening – climate models are complex and accurate long-term predictions are extremely difficult to make – there is no doubt that we need to take action to prevent changes going beyond a point where they cause significantly more damage to human habitats.

At the same time, there has been a lot of publicity in recent years about plastic waste. Plastics are incredibly useful materials, whether in healthcare or food preparation. They have saved many lives. In the right place and properly used and disposed of, they have been a boon. But when plastics end up in the ocean, they can cause considerable problems for wildlife. A host of celebrity efforts in the media to reduce plastic waste has resulted in a lot of confusion. You will often hear and see media reports that conflate climate change and plastic waste.

The surprising thing about plastic waste, given its recent demonization, is that in one way it is good for the environment. Plastics take a long time to degrade – to break down to their component elements and simpler compounds. While a plastic remains intact, just like a tree it is locking away carbon that could otherwise be released into the atmosphere, making climate change worse. If we compare plastic packaging, for instance, and the biodegradable packing now favoured by companies wanting to appear green, both will cause some greenhouse gas emissions in production, but the biodegradable packaging is far worse for climate change as much of its carbon will end up as carbon dioxide, rather than safely sequestered in the earth. Using biodegradable packaging is greenwash – trying to look green when you really aren't – as far as climate change is concerned.

It's good to reuse plastics, and to dispose of them in an environmentally friendly way when they are no longer usable. We certainly don't want them ending up in the ocean. But moving away from plastics to biodegradable alternatives is worse for the most pressing environmental issue, preventing climate change.

— 12 —

ATOMS ARE LIKE MINIATURE SOLAR SYSTEMS

The picture of an atom as a miniature equivalent of a solar system is both visually iconic and appealing to our sense of order. Ask a graphic artist to produce a symbolic representation of an atom and it will almost inevitably look like a star with planets flying around it – look, for example, at the logo of the International Atomic Energy Authority. We know that a solar system consists of a massive star with relatively small planets orbiting around it at a distance. And an atom has a massive nucleus with relatively small electrons orbiting it at a distance. There's a beautiful kind of symmetry to this picture. Unfortunately, it is completely wrong, something that was realized soon after it was discovered that atoms had an internal structure.

The original concept of an atom was as the smallest possible component of a piece of matter. The word comes from the

Greek *atomos* – literally meaning uncuttable. When you take a substance and cut it into smaller and smaller pieces, eventually you can cut it no more, and what you have left is atoms.

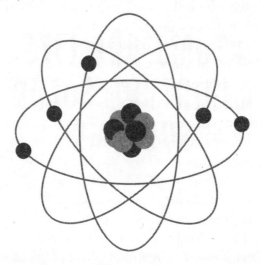

This picture started to fall apart at the start of the twentieth century when the New Zealand physicist Ernest Rutherford and his team, working in Manchester, showed that most of the mass of an atom was in a central nucleus. (Rutherford borrowed the term 'nucleus' from biology, where it was already in use for the central component of a complex cell.) It was also discovered that atoms contained tiny electrically charged particles called electrons. The electrons were negatively electrically charged, and the nucleus positively charged, cancelling each other out. But what wasn't clear was the internal structure of the atom.

Before Rutherford's definitive experiment, the discoverer of the electron, Cambridge-based English physicist J. J. Thomson, had proposed a 'plum pudding' model, where electrons were spread through a sort of positively charged goo, like the raisins in a Christmas pudding. (This term has caused considerable confusion, as there aren't any plums in a Christmas pudding – plum here is just an ancient term for raisins.) But the existence of a positively charged nucleus required the electrons to be somewhere in the outer parts of the atom.

It was at this point that those who were trying to deduce the structure of the atom, notably the Danish physicist Niels Bohr, briefly dallied with the possibility of a parallel with a solar system. In our solar system, planets like Earth are falling in towards the sun, attracted by the force of gravity. But they are also moving at right angles to that direction[3], as a result of which, although they are falling, they keep missing. It's pretty much the definition of being in orbit. So why not imagine that the electrons, attracted to the nucleus by the force of electromagnetism, were also in orbit?

The problem is that when an electrically charged particle is accelerated, it loses energy in the form of light. That's how a radio transmitter's aerial works. Electrons are accelerated up and down the aerial and give off light in the radio part of the spectrum. If the electrons weren't being driven by the transmitter, they would soon lose their energy. But if electrons were in orbit they would also lose energy this way because they are constantly changing direction, and changing direction is a form of acceleration. Electrons would soon plunge into the nucleus. All atoms would self-destruct, pretty much instantly.

---

3. The motion of the planets in orbit around the sun, at right angles to gravitational attraction, arises from the natural tendency of contracting clouds of gas and dust to rotate, due to the asymmetric distribution of matter.

Bohr and the other physicists involved in the development of quantum physics would save the atom by doing away with orbits. Electrons are quantum particles which, unlike a planet, don't have specific locations unless they interact with something else. The rest of the time the electrons exist as clouds of probability, surrounding the nucleus. And their quantum nature means that they can only lose (or gain) energy in chunks – quanta of light known as photons. This means that electrons can't plunge into the nucleus – they can only jump from one cloud to another in what is known as a quantum leap. These clouds are known as orbitals to avoid (or possibly to cause) confusion with orbits.

It might be artistically convenient to portray an atom visually as a miniature solar system, but that picture bears no resemblance to the real thing.

— 13 —

NOTHING CAN TRAVEL FASTER THAN LIGHT

It's one of those scientific 'facts' that lurks somewhere in the back of the memory for many of us. There is a universal speed limit, the speed of light, which is around 300,000 kilometres per second (186,000 miles per second). But, as is often the case in science, the reality is more complicated than it first seems.

One complication is a result of the rather fuzzy concept of 'the speed of light'. Light does not have a single fixed speed – its velocity is dependent on the material it is passing through. It is slower, for instance, in glass or water than it is in air, and slower in air than it is in empty space. This is why the effect known as refraction occurs, where the direction of light changes as it passes, for instance, from air to water – the change of direction is a side effect of light's speed changing as the beam moves across the interface at an angle.

While it is true that you can't move a physical object through space faster than the speed of light in a vacuum, there's nothing to stop an object moving faster than light travels in, say, water. (The speed of light in a vacuum, incidentally, is exactly 299,792,458 metres per second, as a metre is now defined as the distance light travels in 1/299792458th of a second in a vacuum. It's a shame they didn't decide on the round 300 million.) When an object does travel faster than light in a medium, the result is an optical equivalent of a sonic boom, the build-up of pressure waves produced by a plane travelling faster than sound that produces a loud bang.

This effect can be seen in the so-called Cherenkov radiation that causes the eerie blue glow seen around a nuclear reactor that is housed under water. Extremely high-speed electrons are produced by the reactor, which travel through the water faster than the approximately 226,000 kilometres per second (140,000 miles per second) that light manages in this medium. In the process, the passing electrons stimulate the energy of other electrons within water molecules, which then drop back in energy, giving off the blue light.

There is also a subtle aspect of the speed limit that arises from Einstein's special theory of relativity. This theory shows that time and space are not separate things, but part of an interrelated whole. When something moves through space at speed, relativistic effects mean that, to an outside observer, time slows down for that object, the object shrinks in the direction of travel and its mass increases. The factor by which this change occurs depends on the speed of light – if an object were accelerated to the speed of light, its mass as measured by an external observer would become infinite. (As far as an observer on the object is concerned it isn't moving, so there is no effect – that's why the theory is relativistic and rather mind-boggling.)

This impossibility of accelerating with a near-infinite mass means that we can't observe an object moving through space that accelerates past the speed of light. In principle, however, an object that always moved faster than light could exist: such a hypothetical particle has even been given a name – a tachyon. But there is no evidence that tachyons are real. There is, though, a different way to produce effective faster-than-light movement – by changing space itself.

The limitations of special relativity don't apply if space is expanded or contracted. Imagine, for example, we draw two dots on the surface of a partly inflated balloon, then blow up the balloon some more. The dots move away from each other – yet they are both still on exactly the same bits of the balloon where they first started. The dots haven't moved through 'balloon space'. It's the balloon itself that changed. Similarly, we know that space can expand or contract. It's thought that the early expansion of the universe was far faster than the speed of light, made possible because it was space changing rather than objects moving. Mexican physicist Miguel Aclubierre has even speculated about the possibility of having a warp drive, not unlike that of the USS *Enterprise*, that would shrink space in front of a spaceship and compress it behind, moving the vessel forward without it travelling through space.

A similar small-scale effect has been demonstrated when light itself is sent faster than light in so-called superluminal experiments. Here, an ability of quantum particles called tunnelling is employed. Because quantum particles, such as photons of light, don't have a definite location until they interact with something else, existing as a spread of probabilities for their location, they are able to jump straight through a barrier that should stop them. The time taken in getting through the barrier is far less than the time that light would take to travel that distance – photons have been measured travelling at over four times the speed of light, and have even been used to carry a signal of superluminal music. (Visit *bit.ly/supermozart* to hear music that has travelled at 4.7 times the speed of light.)

— 14 —

BLOOD IS RED BECAUSE OF THE IRON IN IT

I can't honestly recall much biology from school, but I do remember being told that blood is red because of the iron it contains. This idea seems to make sense. Although iron is a silvery metal, a number of iron compounds, notably iron oxide (better known as rust) are orange or red. And blood contains a chemical compound known as haemoglobin: that 'haem' part of the name comes from the ancient Greek for blood and is often used to describe iron-based materials, such as haematite, the mineralogical name for a common iron ore.

What's more, there is a form of anaemia known as iron-deficiency anaemia, which can occur due to blood loss or in pregnancy. The condition is discovered using a blood test, and treated with 'iron tablets' – thankfully not literal chunks of iron, but pills containing iron sulfate, a compound of iron that helps someone with the condition to recover their iron levels.

When you zoom in to the molecular structure of haemoglobin, there, indeed, is the expected iron. Haemoglobin is a protein – one of thousands of worker compounds playing essential roles in living organisms. Originally, the organic compound was called haematoglobulin (which literally translates to something like a 'little blob of blood'), but this proved a bit of a mouthful. It's not just humans that use this protein as an oxygen carrier – so does every other vertebrate other than fish.

Apart from water, haemoglobin is the prime constituent of the red blood cells that have the essential job of carrying oxygen around the body (they also carry unwanted carbon dioxide to dispose of it). These tiny cells, shaped a little like miniature dried apricots, pass around your body in around twenty seconds and last for about four months before being replaced.

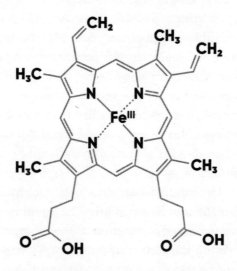

Structure of heme unit with iron (Fe) at its core.

Every molecule of haemoglobin contains four iron atoms, each being part of a relatively tiny 'heme' unit that, with its surrounding organic material, makes up a structure called porphyrin. Again, the word's Greek origin can help us here – the 'porphyr' part is derived from the ancient Greek name of murex snails, the shellfish used to produce Tyrian purple dye once favoured by emperors. What's interesting about porphyrin is that its colour changes depending on its shape as it is the interaction of light with the molecular structure that determines haemoglobin's apparent colour – and its shape is altered when it is doing its job of carrying oxygen. The shape modification transforms the relatively dark red of blood to a brighter red that shows it's got oxygen on board. But neither of these colours closely resembles the orange-red hue of rust.

It is the colour-changing habit of porphyrin that gives the telltale red flush on the skin found in a victim of carbon monoxide poisoning. Carbon monoxide is an insidious, odourless, invisible gas that is particularly good at binding to porphyrin – so good that the porphyrin prefers to latch onto it rather than oxygen, meaning that the body doesn't get the oxygen that it needs. The shape change with carbon monoxide rather than oxygen attached results in that startling skin colouration. This doesn't occur when haemoglobin is carrying carbon dioxide as that attaches to a different part of the heme groups.

So, blood does contain iron – and it is indeed the part of blood that has the iron in it that gives the liquid its red colour – but it is not the iron that produces the red colour, unlike the iron-red of rust or the surface of the planet Mars, named Ares by the ancient Greeks as the planet of the god of war because of what they saw as its blood-like colouring.

— 15 —

HUMANS ONLY USE 10 PER CENT OF THEIR BRAINPOWER

I started my writing career as an author of books on business creativity, and I am ashamed to confess that in one of those books, when encouraging readers to use techniques to give their brains a workout, I blithely informed them that we typically only use 10 per cent of our brains, leaving a huge untapped resource. Unfortunately, this widely held belief has no basis in reality.

Those attempting to sell products that 'enhance your brainpower' or provide brain training often assert that we do under use the capacities of our brains – and they often make passing reference to scientific studies that appear to demonstrate this. The earliest clear example of a study that could at least suggest that we use our brains less effectively than we might was undertaken by a pair of psychologists at Harvard University at the end of the nineteenth century.

Strictly speaking, the American scientists William James (brother of the novelist Henry James) and Boris Sidis did not say that humans fail to use most of their brains, but suggested rather that we don't generally think as effectively as we could, which is hard to deny. But indirect references to their Harvard study seem to have accidentally shifted the claim to the idea that there are big chunks of the brain that sit around doing very little for most of the time.

An influence in this shift was probably the discovery that the brain is not a homogeneous organ. As a result of studying victims of accident and disease, who had parts of their brain damaged or destroyed, it was discovered that different areas in the brain (or, more strictly, different volumes, as it's a three-dimensional structure) are responsible for the many roles that the brain undertakes, from creating and retrieving memories to controlling the parts of the body, decision-making and building a picture of the world around us from sensory inputs.

We now have devices such as MRI scanners capable of detecting activity across all the parts of a functioning brain, so rather than attempt to guess what's happened when bits are knocked out, scientists can study exactly which parts of the brain fire up when we undertake different activities. Although there are relative differences in levels of intensity, these scans show us that the vast majority of the brain remains active whatever our activity. It's true that parts of the brain help coordinate different roles, and some parts are more active than others at different times, but brain function still relies on interaction with a far wider set of the other parts than was originally assumed.

A good example of location-based theory that has now been superseded by better information from scans is the left-brain, right–brain split. For a long time, it was thought that the left brain was associated with logical, structured thought – it was the cold but rational and scientific side of the brain – while the right brain was in charge of art and colour, emotion and creativity. This left/right concept was based on the observation that the brain is, in fact, two almost separate parts, only joined by a large bundle of nerves called the corpus callosum. However, while it's true that our brains do handle some of these mental activities in slightly different ways, modern brain scans have proved that the left/right split is a hopeless oversimplification.

Those who were enthusiastic about the idea that only 10 per cent of the brain was typically at work made use of it to suggest that superhuman abilities could be tapped into if only we could make use of the inactive parts. Some suggested that mental abilities such as telepathy and telekinesis were waiting for us if only those unused brain sections could be activated. But with a bit of thought, it all seems very unlikely. If so much of the brain were hardly used, it's unlikely that this complex organ, which takes up about 20 per cent of our total energy use, would not have lost much of its complexity over time as humans evolved.

It's not that we can't improve our ability to think rationally or creatively. To that extent, we certainly don't use our brains to their full capability. But it's nothing to do with large parts of the brain sitting around, doing nothing, when they could be usefully employed.

— 16 —

A COIN DROPPED FROM THE EMPIRE STATE BUILDING COULD KILL YOU

The Empire State Building doesn't even make it into the fifty tallest buildings in the world anymore. At the time of writing, it's only the seventh tallest in New York City. Yet the combination of it topping the world's building height charts for a lengthy period between 1931 and 1972 and its appearance in over two hundred and fifty films, starting with the iconic scene in *King Kong*, mean that it remains a visual and emotional touchpoint for anything requiring a significant measure of height.

It is likely that the idea that a coin dropped from the top of the building would be deadly to someone on the sidewalk below arose when the Empire State Building still held its crown as the tallest building in the world. It seems quite a reasonable assertion. After all, many coins are heavier than bullets, and a bullet can cause terrible damage. The amount of oomph with which something hits a target is measured by its momentum – the mass of the object times the velocity at which it travels. Current British coins weigh between 3.5 and 12 grams (0.12 to 0.4 ounces), while US coins range from 2.3 to 11.3 grams (0.08 to 0.4 ounces). Not a lot. But the deadly capabilities of a falling coin rest on the assumption that it can get up to a considerable speed when dropped from the height of the Empire State Building.

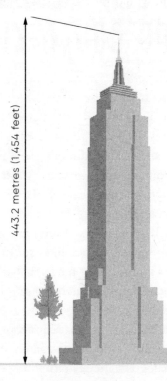

The exact height involved is a little vague as the Empire State Building is topped with a radio mast and a high pointy roof (added to the original design to make sure it was taller than the rival Chrysler Building) – but even if the coin thrower were to climb to the upper parts King Kong-style, it would be extremely difficult to throw a coin and manage to get it over the edge of the building, so we probably have to assume that the coin would be dropped from the observation deck 320 metres (1,250 feet) above the pavement below.

How fast, then, would the coin be going when it reached its potential victim? The starting point is the acceleration due to gravity. Although the Earth's gravitational pull drops off as you move away from the planet's surface, the building's height has a trivial effect. Gravity acts as if the planet's mass were concentrated at its centre. When we stand on the Earth's surface, we are on average 6,371 kilometres (3,959 miles) from the centre. There is not going to be much difference between 6,371 kilometres and 6,371.32 kilometres.

The acceleration due to gravity at the Earth's surface is 9.8 metres (32 feet) per second, per second. That's to say, after 1 second you are travelling at 9.8 metres (32 feet) per second, after 2 seconds at 19.6 metres (64 feet) per second and so on. Working out the speed this implies would be reached in a fall of 320 metres is not totally trivial, but there are plenty of calculators out there. If nothing else were involved, a coin would take 8 seconds to make the drop and would arrive travelling at around 79 metres (259 feet) per second. Going for a coin of 10 grams (0.35 ounces), this would give a momentum of 79 × 0.01 = 0.79 kilogram metres per second. To put that into context, a handgun bullet can have a momentum of around 450 × 0.007 = 3.15 kilogram metres per second – around four times as much. (Converting this to non-metric units is messy and probably no more meaningful.)

In practice, though, there's another consideration. The air. We tend to ignore it, but falling objects are slowed down by the atmosphere's resistance to objects moving through it. As a result, any object has a 'terminal velocity' – the fastest speed it will fall through air, depending on how much resistance the profile of the object puts up. (This is why a parachute, with much more surface area presented to the air, slows a person's descent, compared to falling without one.)

For a typical person, that terminal velocity is around 55 metres per second (180 feet per second) belly down – for a coin it's likely to be around 28 metres per second (92 feet per second), which would take its momentum down to around a twelfth of that of a bullet.

Being hit by a coin falling from the Empire State Building would certainly be unpleasant – but it won't kill you. The TV show *MythBusters* created a special gun to fire a coin at the appropriate rate and showed that it was survivable (please don't try this at home). And a more comparable experiment from around 2007 showed that coins were even less dangerous than the *MythBusters* experiment suggested.

Louis Bloomfield, a physics professor from the University of Virginia, devised an experiment that automatically dropped a whole cache of coins from a weather balloon, high enough up for them to reach terminal velocity. He claimed that they didn't hurt, feeling like the impact of heavy raindrops. Bloomfield used one cent coins, lighter than the weight used above, but also found an additional slowing factor. The coins' unstable fluttering tended to reduce their speed to as little as 11 metres per second.

— 17 —

SUGAR MAKES CHILDREN HYPER

It's a common feature of comedy shows featuring children, from *The Simpsons* to *Modern Family*, that at some point a child will consume too much sugar and start practically bouncing off the walls, they become so energized. They are described as being 'hyperactive' – hyper for short.

In the real world, there is no basis for this commonly held belief. There have been at least twelve good studies that show no connection between sugar consumption and changes in behaviour. To be good quality, a study like this has to be 'double blind'. This means that the children taking part don't know if they have eaten sugar or not – some get a 'placebo' that contains no sugar but tastes the same – and even the experimenters don't know which of the children had which version of the treat, as otherwise they could be biased in their interpretation of the results. There was also no effect in children who had been diagnosed with ADHD, or whose parents thought that their kids were strongly affected by sugar.

What these studies show is that the biggest influence on the reported outcome from everyday life of eating sugar was the expectations of parents. If adults think that their children have consumed sugar, they see more hyperactive behaviour because that is what they expect. These expectations don't come from nowhere, but they highlight a classic problem in interpreting scientific results: confusing correlation and causation.

Correlation is when two things vary in the same way over time or place. Our assumption tends to be that this relationship is causal – so factor A causes factor B. But it could be that rather than A causing B, B causes A, or that both effects are caused by a third thing. Equally, with enough sources of data, there will be plenty of correlations that have no cause at all – the way the two measures track each other is pure coincidence. There is so much data now available in the world that there are many such spurious correlations. A website with graphs of them has been set up, including the fact that eating cheese seems to cause people to die by being tangled in their bedsheets, and the rate at which margarine is consumed appears to be responsible for the rate of divorce in the US state of Maine.[4]

It is likely that the idea that sugar causes hyperactivity is a misinterpretation of the likelihood that both consuming sugar and being hyperactive are caused by a third factor. If you think of circumstances where children are likely to have more sugar than usual, they are often at birthday parties, celebrations or given treats for achievements. In all these cases, the reason more sugar is being consumed is also likely to cause more excited behaviour. Children get excited because they are, say, at a birthday party,

---

4. See tylervigen.com/spurious-correlations

rather than because they are eating sugar at the party.

This isn't to say, of course, that overconsumption of sugar is a good thing. There are plenty of reasons, from tooth decay to increasing the risk of diabetes and heart disease, for keeping our sugar consumption down. But avoiding hyperactivity is not one of them.

If it isn't the sugar, the colouring used in sweets is also sometimes blamed for hyperactivity – notably the yellow dye, tartrazine, also known as E102. The fact that it's an E number is enough for some of us to be suspicious, though in fact the E numbering system – the European Union's food additive regime – includes all additives good and bad. E300, for example, is vitamin C.

Tartrazine used to be widely employed as a food colouring because it is both cheap and very stable when compared with many natural colourings. But it got a bad name from a pair of studies linking it to childhood hyperactivity. Although there has been a large-scale withdrawal of the use of tartrazine, there have been contradictory results from other research – unlike the sugar studies, though, there was not sufficient evidence to be sure of the relationship between the tartrazine and the behaviour. The best evidence suggests that if there is an effect, it isn't due to tartrazine alone, but it's not surprising that the colouring has largely been dropped just in case it could be a contributory factor.

— 18 —

IN THE MIDDLE AGES EVERYONE THOUGHT THE EARTH WAS FLAT

There is a widespread belief that the people of the Middle Ages were scientifically backward. The story goes that they were kept down by the oppression of religion: scientific ignorance was the norm and, for example, the Earth was believed to be flat. This simply isn't true. Not only would most educated people have known that the Earth was spherical in the Middle Ages, but this was also known all the way back to ancient Greece.

Two pieces of evidence would have suggested to early observers that the Earth was a globe. One was that those who travelled to different parts of the world would see different positions of the stars at night – or even totally different stars. The other is that travellers on the sea would notice distant objects rising over the horizon – and if you went up the mast, you would see your destination before those on deck could, suggesting a curvature to the Earth's surface.

A fine description of this comes up in *Opus Majus*, a book written in 1267 by the English friar Roger Bacon, which forms the proposal for a never-to-be-completed encyclopaedia: 'We know by experience that he who is at the top of the mast can see the port more quickly than a man on the deck of the ship. Therefore it remains that something hinders the vision of the man on the deck of the ship. But there can be nothing except the swelling of the sphere of water.'

We aren't one hundred per cent sure who was first with this observation. Pythagoras used to be given the honour, but there was a tendency historically to attribute far more to Pythagoras than he deserved. We do know, though, that the philosopher Plato, writing around 400 BC, was able to liken the Earth to a ball, while in the third century BC another Greek philosopher, Eratosthenes, made an attempt at measuring the circumference of the Earth's globe by noting the angle of the sun at midday from two different locations and throwing in a bit of geometry. His estimate was around 42,000 kilometres (26,000 miles). This wasn't bad at all, bearing in mind the original definition of the kilometre was 1/10,000th of the distance from the North Pole to the equator.

One possible confusing factor in interpreting what medieval people thought is that some of their best-known maps were not for navigation, but rather used to illustrate conceptual relationships. For example, the wonderful Mappa Mundi (map of the world) in Hereford Cathedral, dating back to around 1300, looks as if it is a map of a flat Earth with Jerusalem at its centre – but this is to show the city's place as a spiritual centre, not representative geography. For that matter, concepts of perspective in drawings and paintings, or accurate projection of maps onto a flat surface, had yet to be developed.

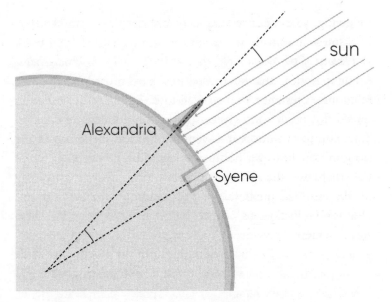

Eratosthenes used the difference between the angle of the sun at two locations to estimate the Earth's circumference.

The story of 'ignorant flat Earther medieval folk' was an intentional construct, dreamed up in the nineteenth century in an attempt to demonstrate the irrationality of religion. One of the main sources seems to have been a book called *A History of the Life and Voyages of Christopher Columbus*, published by the American writer Washington Irving in 1828. In TV form, this would be described as a drama documentary – there was plenty of fictional content in the book, including an invented dispute between Columbus and the Catholic Church on the spherical nature of the Earth, used to add drama.

A number of early historians of science contributed to the made-up attempts to display medieval ignorance over the following years. For example, American historian Andrew Dickson White made it a significant feature of his 1896 work *A History of the Warfare of Science with Theology in Christendom* – but it is clear that such attempts were motivated by anti-religious sentiment rather than historical evidence. Those who spread this myth were concerned by religious fundamentalist objection to evolution and were prepared to make use of any ammunition, however fictional it might be.

Perhaps we should leave the discussion with a reflection on the assumed medieval viewpoint in the opening of John Donne's Holy Sonnet VII, published in 1633: 'At the round earth's imagin'd corners …'

— 19 —

GLASS IS A VISCOUS LIQUID

Sometimes a science myth emerges from an attempt to produce a fascinating scientific factoid that will cause surprise to your audience, only for it to turn out that the most surprising aspect of the 'fact' is that it wasn't true in the first place. This is the case with the idea that glass is a viscous liquid – a gooey fluid that is very slow to flow.

This was a widespread 'fact' that was commonly mentioned in the twentieth century. Of course, the suggestion was never that glass was an easy-flowing liquid like water, which runs away very quickly, but rather that it was so thick and gooey that it would take centuries to produce any detectable movement.

Although this theory seems highly unlikely, there was a reason for coming up with it. If you take a medieval piece of window glass out of its frame, you will find that it is usually a bit thicker at the bottom than it is at the top. It seems reasonable to assume, then, that the glass has run down under the pull of gravity. This would be a very slow process, but one that was observable after hundreds of years had passed.

However, this turns out to be a misguided explanation for a phenomenon with a totally different cause. Until technology was developed to produce sheets of glass with even thickness, starting in the nineteenth century and coming to full fruition in Pilkington's float glass process from the 1950s, window glass was made in small sections that were rarely of consistent thickness. When a glazier put these panes into a window, they would always put them in with the thickest part at the bottom, as this was the most stable way of fitting them. The glass hadn't run down over centuries – it was always that way.

Admittedly, glass is not like a lot of other solids. A solid is distinguished from a liquid because electromagnetic bonds hold the atoms roughly in place – though they do still vibrate. In many solids, these bonds form regular patterns. This can be seen in the structure of crystals and applies both to the familiar translucent crystals such as gemstones and also to less elegant-looking solids such as the graphite that is used as pencil lead.[5] Glass, however, is an 'amorphous' solid. It does not have a regular internal structure; instead, the atoms are linked in a random mix of ways.

This amorphous structure tends to form when a liquid is cooled very quickly and can be seen occurring in nature when magma – molten rock – drops rapidly in temperature. Most of the glass we come across in everyday life consists of variants made from silica, which is another name for silicon dioxide, an extremely common compound in nature, turning up as quartz and as the biggest component of sand.

Although glass is not a viscous liquid after all, there are substances that really are liquids yet still flow extremely slowly. The best known of these is the thick petroleum derivative known as bitumen, asphalt, or pitch – the black stuff used in making tarmacadam roads. A famous experiment involving this substance was started at the University of Queensland in Australia in 1927 and is still underway. It consists of a glass funnel full of pitch, suspended over a beaker, all under a glass dome. In the time that the experiment has been running, just nine drops have fallen from the funnel; the last one (at the time of writing) was in April 2014. The tenth drop is expected to fall sometime around the end of the 2020s.

If watching paint dry is a bit racy for your liking, the university provides a live video feed to keep an eye on the pitch drop experiment at www.thetenthwatch.com.

---

5. Pencil lead is carbon, not lead. The confusion appears to have arisen because the lead ore galena is a shiny black crystal that looks very like natural graphite crystals.

— 20 —

MORE HUMANS ARE ALIVE NOW THAN HAVE EVER LIVED BEFORE

There's a long-standing 'fact', dating back at least to the 1970s and still bandied around, that there are more people alive now than have lived in the entire past of humanity. It is certainly easy to be shocked by the scale of the human population. At the time of writing, the headcount stands at around 7.8 billion people. That's the kind of number that is very difficult to get your head around. If you started counting people, and ticked off one a second, it would take you 247 years to number them all (by which time all the people you started with would be dead).

What certainly is true is that the Earth's human population has shot up. As we saw in looking at this data in chapter 8, at the start of the nineteenth century there were about a billion of us. It took all the way to the 1920s to reach 2 billion. In 1960 we hit 3 billion – this would double in the forty years to 2000. If population growth went on at this rate then the number of humans would become truly vast, to the extent that we would run out of resources – but in reality, thankfully, growth is slowing. It is estimated that if current trends continue, the population will level off at between 10 and 12 billion before the end of this century, after which it will decline.

Because growth has been so fast, it's easy to imagine that the much smaller human populations that were typical in the past could be outnumbered by all those now alive – that the living would outnumber the dead. But it's just not true if we look at the size of the population as a whole. Of course, there are some subsets of the population where we can indeed say there are more alive now than have ever lived before. For example, it has been estimated that around 90 per cent of all the scientists who have ever lived are currently with us today. Far more individuals are scientists now than at any point in history. But it's different when we consider the size of the entire population.

Bear in mind that the species *Homo sapiens* has probably been around for 300,000 years. Initially, there would have been relatively few in the population. And there have been times, as well, since humans evolved when the Earth's climate has changed drastically, and the total human population alive plummeted and was probably measured in thousands. But even so, it has been estimated that around 120 billion humans have lived since *Homo sapiens* first emerged as a species.

This is, of course, a guesstimate, made by an organization called the Population Reference Bureau (they actually came up with a figure of 117 billion, but that seems far too precise for a guess to me). Even if this figure is a bit of an overestimate, it is very hard to see how we could get the total figure of humans who have ever lived down to less than 15.6 billion. This is the maximum possible if we add together the current 7.8 billion and a smaller number for the humans who came before.

– 21 –

ADA LOVELACE WAS THE FIRST PROGRAMMER

One of the great steps forward in the history of science and technology in recent years is the acknowledgement of the many women who have contributed to that history, but who in the past were overlooked. One of the leading lights of these was Ada Lovelace.

Born August Ada Byron, she was the only child of the infamous poet Lord Byron and his wife Anne Isabella Byron. With a strong interest in mathematics from an early age (Ada is often described as a mathematician, but was more a gifted amateur, as was common among aristocrats of the time with an interest in science or maths), she was educated to near undergraduate level in the subject and continued to follow mathematical developments, but she never worked in the field.

In 1835, she married William King, at the time Lord King. Soon after the marriage, thanks to Lovelace's aristocratic heritage, King became the Earl of Lovelace, as a result of which Ada became Ada King, Countess of Lovelace, by which name she is commonly known. (Confusingly, the wife of an earl is a countess. This is because the Normans in England kept the Anglo-Saxon title 'Eorl', later spelled 'Earl', rather than adopting the French title 'Count', probably to avoid ribaldry from the conquered Anglo-Saxons.)

Lovelace had been a close acquaintance of Charles Babbage, the inventor of two mechanical computing machines, neither of which was completed in their lifetimes. Babbage's Difference Engine was a sophisticated mechanical calculator, a working model of which was built at the Science Museum in London in the 1990s. But of far more interest to modern eyes was Babbage's Analytical Engine. This was much closer to being a mechanical programmable computer than the Difference Engine. However, it is now thought that the Analytical Engine could not have been built at the time as it required engineering precision that wasn't then possible to achieve.

Lovelace and Babbage were friendly for a number of years and there was talk of a romance before her marriage to King, though Babbage would not have been considered of high enough social status to ever be acceptable as a husband. The pair had certainly discussed the Analytical Engine and Lovelace went on to provide a translation into English of a paper on the device written in French. This was the work of a little-known Italian military engineer, Luigi Federico Menabrea, who later became more famous as Prime Minister of Italy. Menabrea based his paper on a series of talks Babbage gave in Turin. But Lovelace did not just provide a translation: she tripled the length of the English version, adding a range of speculations on the uses of the engine and examples of what these would involve.

It is on these examples that Lovelace's reputation as the first programmer rests – but there are two problems with this. First, the paper describes algorithms rather than programs. An algorithm is a detailed set of instructions for carrying out a task, which could be used in various ways, one of which is to form the basis for a computer program. The actual program converts the algorithm into the specific format of instructions required to be input into a computer. More importantly, even if it were acceptable to call the examples in the document programs, writing them out did not make Lovelace the first programmer. The paper includes several algorithms that Babbage had detailed in his lectures and had developed years earlier. Only one of the algorithms in Lovelace's notes was original.

This is not to undermine Lovelace's role in spreading awareness of the (never-built) technology and its potential, but in our enthusiasm to present positive female role models from history, it is important that we don't distort reality. There is no doubt that Babbage, not Lovelace, was there first in developing these algorithms that weren't strictly programs anyway. And for a true programmer, we would have to wait many more decades.

— 22 —

BATS ARE BLIND

It's a familiar proverbial saying – as blind as a bat. But lying behind that maxim – used to describe something or someone with poor sight or who can't grasp a concept that's obvious – is a mix of misunderstanding and confusion.

Bats aren't blind. But they thrive in conditions where our own eyesight has difficulties. In particular, bats might have been considered to have had poor eyesight because their flight appears to be erratic – in fact, the bat's rapid changes of direction in mid-flight indicate impressive control and awareness of exactly where they are.

What is true is that bats do not rely on eyesight as much as do other predators that catch flying insects, and this is because of their amazing ability to use echolocation. Although this aptitude makes use of sound, it is a totally different sense from hearing. Echolocation builds a picture of the surroundings from the way that rapid, high-pitched clicks produced by the bat bounce back off objects to be picked up and built into a pattern by the bat's brain.

Bats experience the world around them with the opposite of a sensory limitation such as blindness – instead, echolocation incorporates a whole new sense into their capabilities.

What it's like to be a bat and what the bat is aware of is something American philosopher Thomas Nagel pondered in his 1974 paper 'What is it Like to be a Bat?' Nagel was attempting to explore the nature of consciousness, a subject that remains fraught, whether it is being studied by philosophers or neuroscientists. He suggested that if, for example, a bat is conscious of itself then it experiences what it is like to be a bat. But we (not being bats) can never truly understand what the experience is like for them. We could think about what it would be like to be a human who experienced echolocation, but not what it is to be a bat.

We do, at least, know how echolocation works. The bat produces a series of very high-frequency clicks. Human ears can pick up sounds in a range between about 20 and 20,000 Hz. (Hz stands for hertz, the number of complete cycles a wave goes through in a second.) But bats can produce sounds at over 100,000 Hz and most of their echolocation output is outside our hearing range.

The clicks bounce off objects around the bat and the reflected sounds are picked up by the animal's often prominent ears. This enables them to build up a picture of their surroundings using a combination of the strength of the incoming signal and the differences in timing and strength of what each ear picks up.

This is where our language, built around the human senses, breaks down. I said that the bat 'builds up a picture' – but a picture implies the use of sight. Similarly, while it is known that the bat uses timing and intensity, we shouldn't think of the way that we would measure these with something like sonar. The bat doesn't sit down with a computer and work out the implications of the readings it receives – it perceives its surroundings through echolocation.

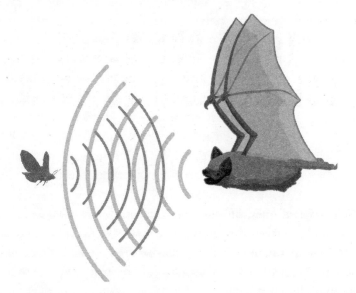

Although it is true, as Nagel suggests, that we can't know what it is to be a bat, we can draw a parallel for what is going on in its head with what we understand from vision. We don't see the world like a camera, capturing a complete picture. Instead, our brains take in a whole range of inputs – shapes, edges, differences in intensity and colour – from the sensors in our eyes and assemble these elements into an artificial construct that we interpret as seeing the world. We can imagine, then, that a bat experiences a similarly constructed but comprehensive 'view' of the world from its echolocation.

Blind? Anything but.

— 23 —

THERE ARE SEVEN COLOURS IN A RAINBOW

One science question you can ask primary school children and adults and get the same answer to is: 'How many colours are there in a rainbow?' There is usually an instant response of 'seven'. Many of us are even taught mnemonics to recall those seven colours, whether it's the historical-sounding British 'Richard Of York Gave Battle In Vain' or the more compact American character Roy G. Biv.

Yet it only takes a moment's thought to realize that this set of seven colours is anything but realistic. Look at an actual rainbow and you may well be only able to see five or six distinct colours — the last three, particularly, are often hard to separate. And as any paint chart will tell you, a far wider palette than just seven colours exists.

The source of the number seven is the illustrious Isaac Newton. Although the instant response to his name is to think of laws of motion and gravity (with added apple), Newton also did a considerable amount of work on light and colour, being one of the first to realize that the white light from the sun is composed of the rainbow spectrum of colours mixed together. It's not a hundred per cent certain how Newton came to that number seven for the colours, but there is a strong suspicion that it was from a parallel with music.

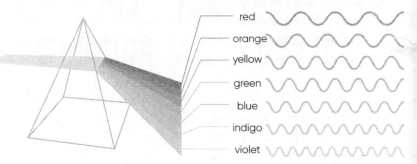

Musical notes are divided into octaves, featuring seven notes (A, B, C, D, E, F and G) before returning to the starting point to make up the eighth note – a range suggested by that name 'octave'. Scientists are often enthusiastic about looking for similarities in the laws of nature – it's thought that Newton felt there ought to be seven colours corresponding to those seven key musical notes.

Interestingly, just a hundred years earlier, Newton would not have been able to pick this particular set, as no colour was called orange at that time. Orange was the name of the fruit, but not a colour; what we would call orange was simply considered a shade of red. This is why some colours in nature that we describe as red – the feathers of a red kite, or the breast of a robin, for example – are clearly orange.

The nature of colour is something that is inseparably mixed up with the way that we detect it. Those colours of the rainbow represent the spread of colours that make up white light – but the rainbow doesn't include every colour. Where, for example, is the brown or the magenta? To understand this, we first need to unpack what colour is. Is it an inherent aspect of an object, or a matter of perception? What we think of as an object's colour is the colour of light that it *doesn't* absorb – and that causes considerable confusion.

Without light, the concept of colour would be meaningless. The colour of light is a measure of the energy of the photons that make it up, which can also be considered as its wavelength or frequency if we think of light as a wave. Higher-energy (shorter-wavelength/higher-frequency) light is at the blue end of the spectrum, and the energy gets lower as we move down through the colours to red.

Any of the colours of visible light can be made up with the three primary colours: red, green and blue. When you look at a phone screen or a TV, each tiny dot (pixel) that makes up the picture has a trio of red, green and blue components, which combine in different intensities to make up a colour. Most modern devices have 256 shades of each of the three primary colours, giving a colour palette of around 16.8 million colours (eat your heart out, Newton).

Yet at junior school we are taught that the primary colours are red, *yellow* and blue. How did green become yellow? It's because of that 'doesn't absorb' bit above. Let's imagine we're looking at a red apple. The white light contains all the colours. But the apple absorbs most of them, sending back only red light. This means that the colours we need to mix for pigments are the reverse of the primary colours, properly called the secondary colours. These are magenta, yellow and cyan.

We're almost there. Magenta, yellow and cyan may be

familiar labels when we're dealing with printer ink cartridges, but at some time in the past it seems to have been decided they were too technical for children to deal with. So magenta morphs into red and cyan into blue – even though they clearly are different colours.

— 24 —

HAIR AND FINGERNAILS KEEP GROWING WHEN YOU DIE

This is a macabre one, which often crops up in classic horror movies. As the creepy music rises to a crescendo, an old coffin is forced open. The desiccated remains inside have long, straggly hair and huge, curly fingernails, because we all know that your hair and fingernails continue to grow after you die. And our source for this idea is not just over-the-top movies. In the 1929 book *All Quiet on the Western Front*, for example, the author Erich Maria Remarque has his main character refer to his late friend's nails growing into obscene corkscrews after death. Except they wouldn't.

When we are alive, our hair grows at a rate of about 10 to 15 millimetres (0.4 to 0.6 inches) each month. Fingernails are slower developers, but still manage to extrude at 3 to 4 millimetres (0.1 to 0.15 inches) during a month.

Hair and nails (everything we can say about fingernails here also applies to toenails, but for some reason they don't get the same attention on the post-mortem growth front) are variants of the same stuff, though clearly the structure of a hair is more flexible than that of a nail. The central material in both is a protein called alpha-keratin. This versatile substance, also found in the outer layer of the skin, provides both protection against abrasion and cold and, in rigid structures such as nails and claws, the ability to grip or rip.

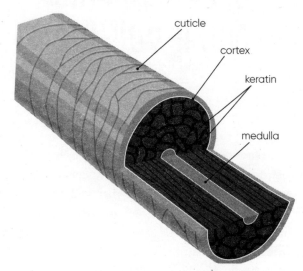

We speak of hair and nails growing, which makes them sound like plants or other living things – but they are no more alive than a piece of plastic or a dribble of saliva. It would be better to think of them as extruded: a construct produced by the living cells at the structure's base, located in depressions in the skin, and pushed out. Hair and nails are dead. This is why adverts that suggest that the manufacturer's product will 'nourish' your hair are immediately flagging themselves as baloney. You can't nourish something that isn't alive.

But you can nourish living cells, like those of your hair roots, which need energy to do their jobs of extruding those tough structures. Typically, this energy is produced by the sugar, glucose, undergoing a combustion reaction with oxygen. Once your body dies, that energy is no longer being provided and the hair and nails stop growing. Not all cells cease functioning after death at the same rate – skin and hair cells do stay alive longer than, say, brain cells – but they will be past the point of no return in a few hours. So where does this widely held belief of post-mortem growth come from? Let's go back to the horror film.

It is true that hair and nails can appear to be longer after death than they were when a corpse was interred. But this is a matter of relative motion. It's not that the hair and nails are pushed out, but rather that the surrounding skin and soft tissue shrink back as they begin to dehydrate. Unlike hair and nails, our bodies contain a lot of water, which we gradually lose after death. The result is that the hair and nails look longer after a time than was the case at death.

This doesn't explain the really impressive growth sometimes portrayed when one of those horror movie coffins is opened – but that is dramatic licence on the part of the effects designer. The appearance of growth in the nails of real corpses is far less spectacular.

— 25 —

HUMAN BLOOD THAT IS LOW IN OXYGEN IS BLUE

Take a look at your veins on the back of your wrists – they are clearly blue. We also refer for historical reasons (of which more in a moment) to aristocratic people as being 'blue-blooded'. Yet no one bleeds blue blood: it's red every time. This isn't true of all species. Arachnids, such as spiders, and a good number of crustaceans and cephalopods, such as squids and octopuses,[6] do have blue blood. This is because instead of using haemoglobin to carry oxygen, these animals employ a chemical known as haemocyanin. But human blood is definitely and always red, though it can come in different shades (see chapter 14 on iron in the blood), being bright red when oxygenated – mostly in the arteries – and dark red when lacking oxygen, as it is in the veins.

---

6. Note: the plural of octopus is not octopi – it is not a conventional Latin word that would have such a plural. As the Edwardian English usage guide Henry Fowler said, octopodes (the strictly accurate plural) is pedantic, while octopi is plain wrong. Stick to octopuses.

We are left, then, with the mystery of those blue veins and the blue-blooded aristocracy. Surprisingly, perhaps, we have blue veins for the same reason that the sky is blue. Everyday light – light from the sun, for example – is white, made up of a whole mix of the colours of the rainbow. Some materials – air, for example, and also the skin – scatter light. This means that the atoms in the material absorb photons of light and then re-emit them in a different direction. But scattering is colour dependent.

The tendency is for the atoms involved to scatter higher-energy, shorter-wavelength light from the blue end of the spectrum, but to let lower-energy, longer-wavelength red light pass through unaffected. So, the blue parts of the sun's light spread across the sky. A more complex process of scattering happens in parts of your body, such as your arm. Where there is no vein, both the blue light and the red light will typically be reflected back from your arm, giving a fairly natural reflection of the colour of your skin. But where there is a vein, the red light continues without being scattered and is absorbed by the haemoglobin in the blood after passing through the colourless vein wall. By contrast, blue light is scattered before it gets as far as the vein and doesn't get absorbed. As a result, the light coming back from the vein is predominantly blue.

What then of our blue-blooded posh people? Biologically, there are few differences between the aristocracy and the rest of us. Historically, aristocrats tended to have more genetic faults than the general population. This was because they only married other aristocrats, and so had a small gene pool, which means that the probability of genetic problems being carried forward into their population was a lot higher than usual. However, they did typically enjoy some environmental differences.

The aristocracy had a more comprehensive diet, so were less susceptible to dietary diseases such as scurvy. And they also avoided sunlight. It's not that there was a tendency for aristocrats to be vampires; rather, it was considered a mark of social status in European aristocracy to have pale skin. This was the result of an aversion to the implications of having a tan. Historically, most of the lower classes laboured outside. To be tanned was a mark of being a menial. The aristocracy avoided work and highlighted this by celebrating pale skins. The paler the skin, the more obvious the blue veins would be. Tanned skin re-emits more of the red hues before they can get as far as the veins. So blue-blooded Europeans were those who lacked a tan.

— 26 —

ORGANIC FOOD IS BETTER FOR YOUR HEALTH

'Organic' is one of those words that we have loaded with more meaning than it can accurately carry. We tend to think of organic food as being healthy and fresh and good. We have to be a little careful when scientific terms are taken over by marketing people, as is the case here. All food is organic, the term properly referring to chemical compounds based on carbon. But in food, the label is applied to foodstuffs that have been produced in a particular way. Being organic in this sense can have impacts on the environment in which the food is produced (good and bad) – but it says little about the healthiness of the food itself.

A number of years ago, I interviewed Helen Browning, then head of the Soil Association, the UK's main organic body. She's a pig farmer and she said to me that the only health claim she made for her organic bacon was that it is better for you than a bag of doughnuts. The fact is, as she was happy to accept, there is very little nutritional difference between organic food and inorganic food.

What, then, is the difference between organic and non-organic food? A cynic would say 'the price' – and it is certainly true that retailers use the organic label to test for price insensitivity in their shoppers. But there are some real differences. Animal welfare tends to be better on organic farms than factory-style farms – though there is no difference between, say, the welfare of organic-reared animals and other free-range animals. There can also be health disbenefits for the animals in the organic regime as organic bodies tend to be supportive of unscientific alternative medical approaches such as homeopathy.

Another big difference is that organic standards restrict fertilizers and pesticides to natural chemicals and products. This can be beneficial for the land, although it has also resulted in dangerous chemicals such as copper sulphate, which was traditionally used as a fungicide, being retained where they would otherwise be banned. But when it comes to health, there is one pesticide-related claim that organic supporters have sometimes made stridently.

In 2001 a representative of the Soil Association wrote in the *Guardian* newspaper: 'You can switch to organic or you could accept that every third mouthful you eat contains poison. Are you up for that?' The suggestion was – and this still regularly comes up – that organic food is better for your health because it retains fewer pesticide residues. However, this is a misrepresentation. In reality, pretty well *every* mouthful you eat contains poisons. This is because plants are in a constant battle with insects and animals that damage them. As a result, they almost all produce natural insecticides and poisons. Which we then consume.

It makes no difference that these are 'natural' poisons. Ricin and botulinus toxin, for example, two of the deadliest poisons known, are both natural. Pesticide residues are far less of a problem for human consumption than natural poisons because you can wash the residues off, which you can't do with the built-in toxins. But this isn't a reason to stop eating fruit and veg.

Practically everything is poisonous with the incorrect dosage – even water. And the quantities of poisons and carcinogens present in food is relatively low. There is a lot more of those natural poisons than there are pesticide residues, even on unwashed non-organic food – but even those quantities are relatively low. There are a few substances – alcohol and coffee, for example – where the carcinogenic effects are quite marked, but even in these they are relatively low. Nonetheless, a single cup of coffee contains more cancer-causing materials than a whole year of consuming residue agrochemicals. So, by all means eat organic food if you wish. But don't expect to be healthier as a result.

— 27 —

YOU ARE BORN WITH ALL THE BRAIN CELLS YOU WILL EVER HAVE

Most of the cells in your body are replaced on a regular basis. Some last longer than others – cells on the stomach lining only survive a few days while red blood cells, for instance, stay around for a few months. Not surprisingly, bones are particularly long lasting, taking around ten years to be renewed. But for a long time it was thought that brain cells were an exception – you were born with a certain number, and those were the cells that you stayed with, less any that died. As you got older, your brain gradually lost its capacity.

The big picture of what is happening to the cells in your body is still amazing. With a few exceptions, the 'you' that you were or will be at age twenty is a totally different 'you' to the age-thirty version. This is reminiscent of the ancient philosophical thought experiment, the ship of Theseus, which dates back to Heraclitus and Plato, but was given an update by seventeenth-century philosophers Thomas Hobbes and John Locke. The original idea involved Theseus gradually replacing all the timber on his ship as various parts of it decayed – is it still the same ship when all the wood has been replaced? What if there's just one original piece left?

Hobbes took the puzzle one step further. Imagine that the parts that are removed are reassembled as a second (rather tatty) ship, piece by piece, over time. Which ship is the original ship – the one that had parts removed or the one assembled from the discarded parts? And if at some point the 'original' label transfers to the reassembled version, at what point does that transference occur?

You could say the same about you as a person. If your body parts are almost all replaced over time, it seems possible that there is some point at which that person becomes someone else. Specifically, though, we tend to think that what makes a person *themselves* is more mental than physical. So, is it no longer the same person if a person loses their memory or suffers deterioration of the brain?

These are not easy questions, fitting better into the field of philosophy than science. However, if we do consider that what makes you *you* primarily goes on in the brain, it is interesting and very relevant if brain cells aren't replaced. Yet modern studies suggest this happens more than was originally thought.

Some parts of the brain definitely do replace cells over time. For example, the hippocampus is a particularly important part of the brain in relation to what we think of as ourselves as it is responsible for processing memories. Strictly speaking we have two of these items in the brain, one for each side. The hippocampus is supposed to look like a seahorse, which gives it its name, though you have to be imaginative to see this. The main job of the hippocampus is to transfer information from short-term to long-term memory – and without long-term memory we lose a lot of what we are.

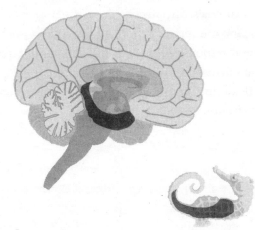

This ongoing development of the structure of the hippocampus with regular replacement of cells is also true of the olfactory bulb, the region of the brain that deals with the sense of smell. Equally, we now know that some parts of the brain gain extra cells when we are young – so it is definitely not the case that we are born with all the brain cells we will ever have.

It isn't an easy task to know if brain cells are being replaced. Scientists can hardly keep opening someone's head and checking. But they have been helped by the fallout from nuclear bomb blasts. These explosions released a radioactive isotope of carbon, carbon-14, into the atmosphere. (Carbon-14 is the isotope used in radiocarbon dating.) Since testing stopped, the amount of this isotope in the atmosphere has been declining. When a cell is formed, its carbon content will reflect the carbon-14 mix in the atmosphere at the time, making it possible to see how the cells are being replaced.

Interestingly, the generation of new brain cells, called neurogenesis, is quite different in adult humans when compared with other mammals. Specifically, there is neurogenesis in the striatum in humans, part of the brain that deals with motor behaviour and response to stimuli. It has been suggested that this is because the striatum has a major role to play in our 'cognitive flexibility' and may reflect our ability to adapt to different environments and circumstances.

– 28 –

LEMMINGS COMMIT MASS SUICIDE

In the early days of computer gaming, a favourite for many was a program called *Lemmings*, where a row of little creatures, constantly on the move, had to be prevented from plunging to their deaths in all kinds of environments. We associate lemmings with a tendency to hurl themselves off cliffs, driven by some kind of mass hysteria, or sheep-like following of a leader. The hold of this idea is so strong that we sometimes say someone is acting like a lemming if they mindlessly put themselves at risk. Yet, in reality, lemmings are not in the habit of committing mass suicide – and the origin of this idea is particularly bizarre.

Lemmings are small rodents about 15 centimetres (6 inches) long, which look like a cross between a mouse and a guinea pig. They live in cold northern regions of Europe, surviving on vegetation. One distinct difference between lemmings and most other small rodents is that they are unusually aggressive for prey animals. They will attack predators – even humans – as their approach to defence. Lemmings also go through large swings in population, typically over a four-year period. Sometimes the population drops to dangerously low levels; at other times, the population soars and lemmings will flood out of an area to find somewhere new to live.

It is this mass migration to find new food resources, combined with the sudden, unexplained drops in population, that may have led to the mistaken idea that lemmings are in the habit of hurling themselves off cliffs, an idea that doesn't seem quite so strange when put alongside one of the earliest ideas about the origins of lemmings. According to a sixteenth-century geographer, they were not unlike hailstones, falling in showers in stormy weather.

It is true that the migration urge is strong in lemmings, and though they are not in the habit of jumping to their deaths, they will attempt to cross bodies of water. Sometimes this can result in many of the rodents drowning, which then has been interpreted as the result of self-destructive urges. This theory was speculated on in 1877 in the magazine *Popular Science*, where the author W. Duppa Crotch suggested the lemmings were attempting to migrate to land that was now lost under the sea. However, the images that would embed the idea of lemming self-destruction into the popular imagination came from the minds of the makers of a Disney documentary.

In the 1950s, Disney made a number of popular wildlife documentaries, including *White Wilderness* in 1958. 'From the top of the world!' the poster proclaimed. 'A fabulous new adventure in exciting entertainment.' Note that last word: this film was more about entertaining storytelling, featuring animals, than accurate natural history. As well as showing the lives of the inevitable polar bears and reindeer, the documentary also told the story of the lemming.

The approach taken in these films was to anthropomorphize the behaviour of the animals, with the narrator describing them as if they had the motivations of human beings. And *White Wilderness* included a sequence featuring lemmings, apparently plunging from a cliff into the sea. To be fair to the production team, the narrator did suggest this was about migration rather than suicide, and that the lemmings had mistaken the ocean for a lake they could cross. But the dramatic visual imagery helped build the suicide myth. Later investigation suggested that the scene was faked. It appears to have been filmed on a river near Calgary, rather than at the edge of the Arctic Ocean, and it has been suggested that the lemmings did not so much leap as be pushed off the edge.

— 29 —

SPIN-OFFS FROM THE SPACE PROGRAMME INCLUDE TEFLON AND VELCRO

Space travel is an expensive activity – and it can be a difficult job to make the case for putting large amount of taxpayers' money into it, especially when it comes to human space flight, where the scientific and practical gains are usually far less important than publicity and the idea of people taking on a new frontier. Indeed, when Amazon billionaire Jeff Bezos made a brief excursion to the edge of space in 2021, he was widely criticized for wasting money, even though this wasn't paid for by taxpayers.

Because of this, NASA and other space authorities have not been shy in putting forward the benefits to be reaped from the many spin–offs of the space programme. The underlying idea is that space flight pushes the boundaries of what is currently possible and, as a result, in trying to solve the problems they face, space scientists and engineers devise new and exciting technology that has gone on to benefit all of us over the years.

Some of these claims come from NASA itself. For example, it rightly points out that memory foam was developed as a result of a NASA contract. But in an attempt to claim something more significant to our everyday lives, NASA has also made the case that PCs would not exist were it not for its need for light and compact electronic computers to fit on space vessels. In reality, the NASA computers were always extremely expensive bespoke devices that bore no resemblance to a PC – it was the development of cheap chip-based processors for traffic-light controllers and other such commercial developments that led the way for a product that required a mass-production approach.

In other cases, though, the link to NASA is mythical (though the agency tends not to go overboard in denying responsibility). The two classic examples of this are the invention of the non-stick material Teflon and of the easy-open fastener Velcro. Though both have been used in space, each predates the existence of NASA entirely – the organization was only formally constituted in 1958. Teflon is a brand name for PTFE – polytetrafluoroethylene. This remarkable substance was developed by accident by American engineer George Plunkett in 1938.

Working on refrigerant gases, Plunkett was worried about the safety of a cylinder he had of tetrafluoroethylene gas. When he turned the on-off valve, nothing came out, but the cylinder was too heavy to be empty. As the gas can be explosive, he tentatively cut through the cylinder behind a blast shield to find a white, slippery substance inside. The gas had polymerized, catalysed by the iron cylinder to form the long chain molecules of PTFE. Initially, PTFE was mostly used by plumbers to seal valves and joints, which would come in useful for space technology – but a French engineer, Marc Grégoire, was inspired by his wife to discover a way to use PTFE to avoid food sticking to her pans. In 1956 he began producing his non-stick cookware under the brand name Tefal.

Meanwhile, shortly after Plunkett's discovery, Swiss engineer George de Mestral had an inspiration when out for a walk. He passed through a field containing burdock plants bearing burrs – seed pods that have evolved to stick to animal fur to spread the seeds, and that also stick to clothing (see above). De Mestral realized that the burr mechanism, which relies on small hooks on the ends of the burr spikes catching on the hairs of passing animals, would make a great quick-release fastener. The development was anything but quick. Velcro was patented over ten years later in 1955 and not produced commercially until the end of the decade – but it would find a wide range of applications, including space engineering.

Most entertaining is a spin-off claim used to suggest NASA extravagance. It is said that the agency wasted millions on developing a ballpoint pen that would work in space, as a standard pen won't write upside down because it relies on gravity to pull the ink down to the ball. This purported NASA 'solution' is contrasted with the approach of the USSR, which is said to have spent nothing on the problem and used pencils instead. While there really was (and still is) a 'space pen' that writes upside down because the ink is pressurized, it was developed by the Fisher pen company at no cost to NASA and it works better in low gravity than pencils, which have a tendency to leave inconvenient fragments of carbon floating weightless in mid-air.

— 30 —

THE BIG BANG THEORY EXPLAINS WHERE THE UNIVERSE CAME FROM

Of all the theories of modern science, the big bang is surely one of the most famous – it's even got its own TV show. To uncover what the theory really does, we need to go back to the 1950s.

At the time, there were two rival theories for how the universe had evolved. Both were based on evidence that the universe was expanding – something that has now been confirmed many times over. The earlier theory, which was developed into the big bang theory, had the universe beginning at a point in time as a sort of 'cosmic egg'. This would imply that the universe (or at least the part of it that we can experience) had a specific moment of creation.

Some astrophysicists were unhappy with this requirement, in part because the atheists among them thought it was too suggestive of the existence of a creator. As Stephen Hawking once put it: 'Many people do not like the idea that time had a beginning, probably because it smacks of divine intervention.' The three originators of the opposing theory, known as the steady-state theory – Fred Hoyle, Thomas Gold and Hermann Bondi – came up with it after a joint expedition in Cambridge to see the anthology supernatural film *Dead of Night*. The very last scene of the film leads directly into the opening scene – it is circular with no beginning or end.

This inspired a picture of the universe that was able to expand yet could keep the same density of matter as had always existed, because matter was constantly being created, preventing it from getting more and more thinly spread. As well as being an excellent astrophysicist, Hoyle was a well-known science communicator: the year after the development of the steady-state theory, he gave a talk on BBC radio. In it, he compared the two theories saying that steady-state 'replaces a hypothesis that lies concealed in the older theories, which assume, as I have already said, that the whole of the matter in the universe was created in one big bang at a particular time in the remote past'.

This was the first use of the term 'big bang', which Hoyle has often been accused of using in a derogatory manner. But on reading the script, it seems more a matter of Hoyle producing a snappy label. Over the next three decades, observations that would enable a distinction between the big bang and steady-state theories began to be made – and came down firmly on the big bang side. Hoyle would modify steady-state theory to match observations (and it ought to be stressed that the big bang theory also had to be modified very significantly too), but steady-state has now lost all support.

There is no doubt, then, that the big bang theory is our best current idea of how the universe got from its early state to its present form. But there is one big hole in the theory – which is something that Hoyle picked up on in that early radio talk. He said: 'On scientific grounds this big bang hypothesis is much the less palatable of the two. For it is an irrational process that cannot be described in scientific terms … it puts the *basic* assumption out of sight where it can never be challenged by a direct appeal to observation.' (Emphasis in the original.)

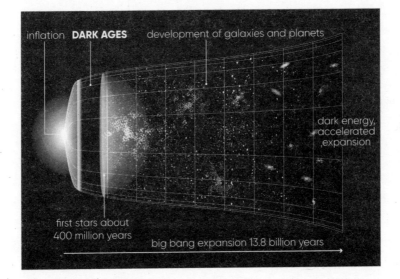

And Hoyle was absolutely right. What the big bang theory fails to do is tell us where everything came from. It begins a tiny fraction of a second after the beginning of space and time – but it tells us nothing about where the universe came from or how it and its natural laws came into being. And it posits a beginning where we can never know about that moment of origin. By comparison, the steady-state theory was much more appealing, because even though it would have been very hard to detect the creation of matter, it was a process that could have been observed.

The steady-state theory may have been wrong – but it is easy to slip from that into binary thinking where we assume that if one of two theories is wrong then the other has to be right. And the problem with the big bang theory is that it can probably never provide scientific answers to the very beginning. It can't tell us where the universe came from.

— 31 —

WE SHOULD RETURN TO THE DIET OF OUR ANCESTORS

For most of the time that there have been human beings in existence – the current best estimate is around three hundred thousand years – we were hunter-gatherers, relying on what we could catch or forage to eat. Around ten to twelve thousand years ago, however, a gradual process of change occurred as humans adopted more settled lifestyles and began to grow crops and domesticate animals. With the introduction of agriculture, our diets underwent a radical change, and nowadays, on a regular basis, we see claims that we should return to what is sometimes called a 'paleo' diet, moving back to something closer to that of hunter-gatherers. This, it is suggested, would reduce our susceptibility to heart disease, diabetes, cancers and more.

Making such a change means abandoning a wide range of familiar foodstuffs: meat from domestic animals, dairy, grains and pulses, sugar and oils. This leaves meat and eggs from wild animals, nuts, seeds, and wild variants of fruit and vegetables. That way, it is argued, we focus on the diet that we evolved to eat and so avoid the unnatural aspects of the agrarian diet.

There are some positives here. Game meat, which is the obvious hunter-gatherer replacement for that of domestic animals, is lower in fat and better for you. What's more, because game (in the UK) has to be shot in the wild rather than farmed and taken to an abattoir, it is far better for animal welfare than farming. But it is considerably more expensive to buy. Cutting out sugar is a no-brainer. But the reality is that this entire 'ancestral diet' is not the best we can have.

One problem is an incorrect assumption: that humans evolved 300,000 years ago and haven't changed since. All species are evolving all the time, responding to changes in their environment, and though the formation of a new species does generally take millions of years, smaller evolutionary changes can be surprisingly quick. Humans have already evolved since the introduction of agriculture. In some parts of the world, for example, we have evolved to benefit from milk as adults, while almost all humans have now undergone an evolutionary change that makes us better at digesting grains than earlier people. We are no longer the species we were 300,000 years ago.

More significantly, just because a diet was prevalent in the past does not mean it is the best diet for a human being to thrive. Evolution is not directed. It doesn't say, 'This is the diet that is available and I will fine-tune people to fit it.' And exactly what we expect from life has also changed. The hunter-gatherer diet, combined with the significant exercise that goes along with that lifestyle, is not a bad one for keeping people healthy to the age where they can breed and keep the species going – which is all that evolution requires. But we now expect long lives after we have had children – something that is not part of the programme.

For that matter, it's not just humans that have evolved – so have the game animals and wild vegetation that is available to us. Nothing stays preserved in aspic in an evolutionary world. We don't know exactly what our ancestors did eat 300,000 years ago, but we can be certain that it wouldn't be the same as today's wild animals and plants.

Some, of course, suggest that we should not eat meat at all – but this becomes particularly problematic if combined with a paleo-style diet because a lot of the alternative sources of, say, protein, such as pulses, are cultivated crops. Humans have always been omnivores and it is almost impossible to get a balanced human diet from foraged plants alone. What's more, eating meat provides a considerable amount of pre-processing of nutrients. Sticking to wild plants means spending many hours a day on eating to get essential nutrients to required levels. It just isn't compatible with living a full modern life.

— 32 —

WATER GOES DOWN THE PLUGHOLE IN DIFFERENT DIRECTIONS EITHER SIDE OF THE EQUATOR

Of all the dubious claims that we meet in this book, this is probably the one that crops up most often in TV documentaries. It is a particular favourite of travel shows when the presenter crosses the equator. We are earnestly told that if you are north of the equator, the water will head down the plughole with a clockwise spin in the mini vortex that sometimes forms in a drain. By contrast, cross over to the south side of the equator and the water will spiral away anticlockwise. Placed right over the equator, water is supposed to head straight down without a vortex at all.

The idea is based on real science – but science that is badly applied. The scientific reason for such an effect is known as the Coriolis force, which is the result of residing on a rotating body. Imagine, for instance, you turn a record player on its side, so the turntable is vertical. Now drop a marble from the middle of the turntable to the edge, so it falls down in a straight line. If you were to watch what happens to the marble from a camera on the turntable, instead of moving in a straight line, its path would be curved because of the turntable's motion. It would appear that a force was pushing the marble sideways, off its straight-line path.

Exactly the same thing applies on the Earth's surface. Because we, as observers, are rotating with the Earth, objects are apparently given a push in a direction sideways to their motion that moves them to the right in the northern hemisphere (producing a clockwise vortex) and to the left in the southern hemisphere.

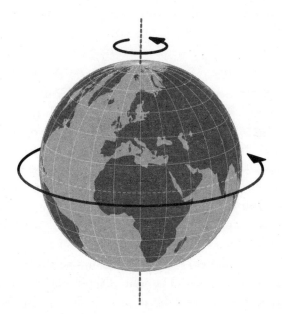

This Coriolis effect is real. It's why, for example, large moving currents of air tend to produce a spiral pattern in the appropriate direction that we label cyclones and anticyclones. However, the problems facing those peering into a sink near the equator are twofold. First, the nearer you are to the equator, the weaker the effect. The presenter usually only takes a few paces either side of the equator, but that is the worst possible place to do the 'experiment'.

The second problem is even more significant. Even if you peer down a plughole a good way from the equator, the Coriolis effect is not the only thing to influence how the water flows. The direction it takes will also be influenced by the location of the tap with respect to the drain and the shape of the basin or other receptacle into (and out of) which the water is flowing. Even the shape of the drain hole can make a difference. The combined impact of these different factors is far greater than the Coriolis force on such a tiny system.

Why, then, do the demonstrations on those TV shows work? Unless you are careful (and these shows rarely are) it's easy to unconsciously make an experiment's outcome reflect your expectations. There was a famous one in which graduate students were given rats to test on mazes. They were given two groups of rats: some, they were told, were especially intelligent, and some were average. Not surprisingly, the brainy rats did better at solving the mazes than the ordinary group. But later it was revealed that the graduate students were the real lab rats. All the rats they were given were identical – there was no special group. But because the students *expected* one group to do better, they gave their 'brainy' rats the benefit of the doubt and generally biased the results in that direction.

That's the generous explanation for those well-behaved vortices. Sadly, it's also true that it is not unheard of for documentary makers to rig things to show the result they want. This applies to water swirling down drains as much as it does to lemmings.

— 33 —

THERE IS NO GRAVITY ON THE INTERNATIONAL SPACE STATION

There's something hypnotic about watching astronauts floating around inside the International Space Station. It's common sense that this happens because they are in 'zero g' – they can hover in space because they have escaped from the Earth and are no longer under the influence of gravity. But the reality is far more interesting. They float around because, along with the space station itself, they are plunging towards Earth. They aren't in zero gravity at all: they are in free fall.

If we ignore the effects of air resistance, how fast something falls under gravity is not influenced by its mass. Big things like the space station accelerate at exactly the same rate as (relatively) small things like people. Someone falling inside a space station that is also falling with the same rate of acceleration will float because of this.

Thankfully, we don't hear of the ISS and its occupants

smashing into the ground. This is because they are moving in a special way that means that, despite falling, they keep missing the Earth. This special way of moving is called an orbit. It involves moving sideways with respect to the Earth's surface at just the right velocity to balance out the acceleration due to the force of gravity. For any particular height above the Earth's surface there is only one speed that the orbiting satellite can move at to stay in orbit.

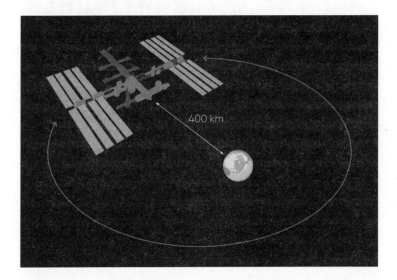

400 km

The ISS orbits relatively close to the Earth: it is around 400 kilometres (250 miles) up. That puts it in space – but close to the planet (officially space starts at the 'Karman line' 100 kilometres – around 62 miles – above the surface). Compare the space station's height with the geostationary satellites used for broadcasting, which maintain position above a particular point on the ground. They have to be nearly a hundred times further from the Earth's surface. Because the Earth is very big, and the ISS is not far from it, the force of gravity affecting the latter is still around 90 per cent of the pull that we feel on the surface. If the space station were on top of a very tall tower rather than in free fall, the astronauts on board would feel almost as heavy as they do on Earth.

The weightless effect of being in free fall can be achieved in ways other than orbiting. For instance, if you are inside a free-falling lift (elevator) you will experience the same ability to float around as do the astronauts – though unfortunately, lacking the ability to avoid a collision that an orbiting satellite has, your experience will be relatively short-lived.

To give people the experience of being in free fall without the fatal finale, special flights are made where a plane climbs to a high point and then dives, leaving those onboard floating for around 25 seconds per manoeuvre. Originally designed to train astronauts, these planes, official known as 'reduced gravity aircraft' and better known as vomit comets, now provide the experience commercially, perhaps most famously giving the late Stephen Hawking, otherwise confined to a wheelchair, a chance to be briefly weightless.

Of course, it is possible to genuinely escape the Earth's gravity. The strength of gravitational attraction drops off with the square of the distance you are away from the centre of the body that is attracting you. Our solar system has some massive objects in it – notably the sun and the giant planets – but a spaceship that got a reasonable distance away from these would be able to provide what was, to all realistic purposes, a true zero gravity experience. Where there is no other source, gravitational effects can be produced through acceleration – the reverse of free fall. Either accelerating a ship or rotating it can generate an equivalent force to being attracted by gravity.

CHIMPANZEES AND GORILLAS ARE OUR ANCESTORS

A number of years ago, I was judging a science communication competition in a secondary school. One of the teams had taken the topic of gorillas. Their graphics asked us to look after the gorillas 'because they are our ancestors'. It was a nice sentiment … but poor science. We are not descended from any of our fellow great apes.

The important distinction here is between having a common ancestor and being in direct line of descent. Our species, *Homo sapiens*, emerged about three hundred thousand years ago. The species we evolved from would have had more of the characteristics that we typically think of as ape-like (bear in mind we *are* apes) than we do. But it would have already been, for example, walking upright.

If you then followed back to the ancestor of that species, and the species before it, and so on, you would eventually get to an animal that is in the family tree of chimpanzees and bonobos too, our nearest relatives among the great apes. Further back in time still, and you would reach our common ancestor with gorillas. You would need to go even further to get to an ancestor we shared with orangutans. As you travel further back up the tree, our joint ancestor would for a while have looked more like a modern monkey – but, again, it wouldn't be one.

Go far enough back in time and you would find a common ancestor that we share with any species of your choice. We would reach, for example, small rodent-like mammals, then precursor species to the mammals, and so on. Eventually you would reach common ancestors with plants. Further still and you would find an ancestor not just of all animals and plants, but even of bacteria. Every species that we have as yet identified, whether living now or in the past, had a single common ancestor – we are all related. Although it is possible that it happened, there is no evidence yet that life began more than once on Earth – there is no living thing that is so alien that we can't find genetic links.

When considering the detail of our specific predecessor species, we are hindered by a lack of information. The fossil record – the remains of organisms that have decayed, preserved when minerals replace soft parts of the body – is sparse. It is extremely unusual for any one organism to be fossilized. The only reason there are a relatively large number of fossils in existence is that there have been so many living things in the 3.5 to 4 billion years that there has been life on Earth. It once used to be fashionable to refer to a 'missing link' between us and an earlier species. In reality, it's more like a missing tree, with just tiny fragments of individual branches randomly preserved.

The way that we can identify predecessors that are close relatives is through DNA. The degree to which DNA alters from species to species gives an indication of closeness of relationship. But DNA itself is limited when it comes to looking far back in time. For example, in 2016, the fragmentary remains of a hominin species known as MRD was discovered. It had some human-like features and dated back around 3.8 million years. The newspapers were full of claims that MRD was our oldest known ancestor. But we simply can't tell for sure. DNA is useless after about 1.5 million years as it deteriorates too much to discover anything useful about relationships. This species was part of the same broad tree as us – but there is no way of knowing if it was in the same branch.

orangutan   gorilla   bonobo   chimpanzee   human

There's one final consideration in calling any great ape our ancestor. Since the ancestors of chimps, for example, and our ancestors split off from their common ancestral line, both species have continued to evolve. We've each had a number of ancestor species since the split. But even since the time that we reached our respective current species, each of these has also evolved. Humans aren't the same as they were 300,000 years ago. Chimps have evolved even more than we have since they first emerged. Evolution never stops.

– 35 –

CHAMELEONS CHANGE COLOUR TO BLEND INTO THE BACKGROUND

Chameleons are amazing animals, most notably for the way that they can change the colour of their skin. We describe people as chameleon-like if they are able to blend into the background – to become invisible thanks to their ability to change the way that they look or behave. So it is, perhaps, a bit of a shock to the system to learn that most chameleons aren't … chameleon-like.

It's not that the colour changing of the animal's skin is a myth. They really are able to do this. But there is no evidence that most species of chameleon use this ability as a means of camouflage – quite the reverse. Chameleons use their colouring as part of their means of communicating with others of their species. The colour change is not to be less visible, but more so. Various animals use sound in their in-species messaging. Many go beyond this. Some insects, for example, use chemicals to communicate, while bees famously use a waggle dance to pass on the location of luscious nectar. Visual communications between animals are also very common – and chameleons take this to the next level.

Typically, bright colours on a chameleon indicate a degree of aggression, while more subtle colouring is a sign that the animal wishes to be cooperative. Chameleons also use their colouring for heat management: lighter colours reflect more incoming infrared, while darker colours absorb it. By lightening their skin when sunlight is intense, the animals can avoid overheating.

The ability to change colour relies on special skin cells called chromatophores. These provide small blocks of colour, rather like the pixels on a phone screen, building up an overall colour pattern. Some species are better than others at the colour-changing trick, with a whole range of colours available, while others are restricted in hue.

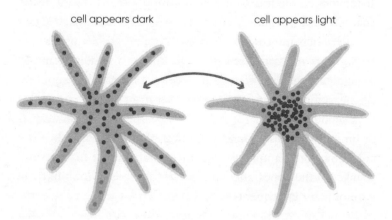

cell appears dark   cell appears light

What's unfortunate in the way that chameleons have (unwittingly) hogged the limelight is that some animals genuinely are good at using changes of colouration to blend in. The most common and skilful of these are flatfish such as plaice, sole and flounder, which live on the seabed and can produce a remarkable match to the pattern of the bottom of the sea, rendering themselves virtually invisible. This is possible because, as well as having chromatophores, the flatfish have light-sensitive patches on the lower side of their body, so can get an impression of the colouration of the seabed before settling on it.

Some octopuses and squids are also effective at using colour for camouflage. The one chameleon that manages to do this to a degree is a variety known as Smith's dwarf chameleon. But this is sufficient of an exception that we probably should call someone who blends in well plaice-like.

Human use of camouflage has tended to be focused on military applications. Here there have been attempts at adaptative colouration, mimicking the supposed ability of the chameleon, but there is often a more subtle approach taken – where it's impossible to make something disappear, the technology makes it look like something else. For example, recognizing that it's pretty much impossible to entirely conceal the heat signature of a tank, one technology instead transforms its heat output so that it looks like a family car. This stealth-by-imitation is also found in nature, though here it is more common, for example, for an insect that doesn't have a sting to take on the appearance of one that does, or to look like the type of plant it frequents.

— 36 —

DIFFERENT PARTS OF YOUR TONGUE RESPOND TO DIFFERENT TASTES

Some of the myths we encounter in science are so widespread that they are even taught in schools. The five senses and the colours of the rainbow are two obvious examples – and the location of the tongue's taste receptors has historically been another. I remember at school doing an 'experiment' involving the different taste areas of a tongue. We mostly didn't manage to match the official map of where taste sensations were located. This wasn't because we were bad experimenters (though we were), but because those taste areas don't exist.

There are five broad taste sensations: sweet, salt, sour, sharp (or bitter), and umami – surely they should have stuck to S-words and gone with savoury? This much is true. However, the old maps of the tongue showing different locations of taste buds for four of the tastes (they were yet to add umami to the palette) are pure fiction. Broadly, the old maps put sweet at the front, bitter at the back, sour at the left and right towards the back, and salty at the left and right near the front. It was even suggested that it is possible to generate those taste sensations by pressure alone on these areas (which was how our failed school experiment was designed).

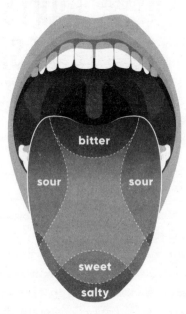

The reality is that a human tongue has between 2,000 and 8,000 taste buds on its surface. These are where the chemical constituents of food, dissolved in saliva, come into contact with taste receptor cells. These biological instruments detect the presence of certain chemicals, which we interpret as a particular taste. So, for example, saltiness is reflected in a relatively high level of sodium ions in the saliva, while sourness involves detectors that can pick up hydrogen ions, which are typical of acids.

Our taste buds sit in the little bumps on the tongue's surface known as papillae. Each bud contains a number of receptor cells to do the job of detecting the appropriate chemicals. These cells work in two broad ways. Most have proteins sticking out from the cell that bind onto the 'taste' chemicals, producing a signal. The definite odd one out is the salt receptor. In many internal structures, the body makes use of membranes that allow, for example, sodium ions to move through them to provide the electrical signalling that organisms depend on to function. A similar 'ion channel' approach is used for the salt taste (and this kind of mechanism may also be involved in sour detection).

To an extent, the 'five tastes' are a bit like the five senses. Other distinctions in food produce what are effectively extra taste sensations. This includes feedback on the texture of what we eat, but also the distinctive 'heat' of peppers and chillies, or the seemingly 'cool' flavour of mint and menthol. Like many of the taste receptors, these are also signalled by cells making use of protein-based detection.

The tongue map is an example of a myth that depends on a misunderstanding of a genuine piece of scientific research. The original work, dating back to the start of the twentieth century, suggested that there was a variation in intensity of taste sensations across the tongue. Somehow this became misunderstood as a variation in the ability to detect different kinds of taste, rather than overall intensity. Exactly who devised the infamous map is not clear, although it is based on a misinterpretation of earlier data by a Harvard psychologist in the 1940s.

As a concept it is not dissimilar to the short-lived Victorian idea of phrenology – the idea that the shape of the skull reflected the capacity of the various functional parts of the brain that lay beneath it. And it is equally fictional.

— 37 —

A BUMBLEBEE'S ABILITY TO FLY DEFIES PHYSICS

Whereas a myth like the tongue map is sometimes still taught in schools, we are more likely to encounter a claim about the impossible aerodynamics of bumblebees on websites and podcasts that try to highlight the limitations of science, wanting us to be amazed by wonders of nature that even science cannot explain.

This viewpoint seems linked to the poet John Keats's dig at Newton and scientists in general in his poem *Lamia*. Keats tells us that these people 'conquer all mysteries by rule and line' and succeed in 'unweaving' the rainbow. It's a petty and, frankly, ignorant viewpoint (sorry, Keats fans). Having an understanding of how something works does nothing to take away its beauty. The rainbow is just as stunning to the scientist – but even more glorious for knowing how it is formed.

What, then, of the mysterious bumblebee? The suggestion in this myth is that the bee's chunky body could not possibly be supported by the flapping of its tiny, thin wings. And even if the wings could manage to get the thing aloft, the bee would need to exert more energy than it consumes to stay in flight. Science just can't explain it, we are told. But in reality, science is perfectly capable of doing so.

The idea here seems to have originated from a sermon, where God was given the responsibility for keeping the bee in flight. But what was needed to explain the bumblebee's impressive flight skills was an understanding of aerodynamics. A bumblebee's wings do not enable it to glide, or to fly with a simple flapping motion like a bird. A closer analogy is a helicopter – the bee's wings move at high speed through curves that produce vortices – spinning columns of air that increase the lift it experiences sufficiently to deal with its plump, but very light, body.

To do this does not require a huge amount of energy – there is no need for energy to magically appear from nowhere to keep the bee in flight. If a sermon writer really wanted to head down that route, an apparently better candidate for defying physics would be the kangaroo. This genuinely does make use of more energy in its vigorous hops than it consumes in food. But, just like understanding the bumblebee, we don't need to call on divine intervention to explain the kangaroo's abilities.

The secret to the kangaroo's apparent defiance of the laws of physics is that we aren't taking into account everything that happens when the creature jumps. A useful parallel is a bouncy ball. When you drop the right kind of ball onto the floor it bounces up again – some, so-called superballs, can bounce nearly as high as they were originally dropped from. The ball doesn't have some secret source of energy. Instead, when it hits the ground, it absorbs the energy of collision and then releases it again, pushing the ball away from the floor without any extra energy added to the system. Something less bouncy – a bag of flour, say – converts pretty well all of the energy of its collision with the floor into noise, heat and disruption of the bonds in the paper as it bursts open.

The kangaroo is more like the rubber ball than the bag of flour. As the animal lands, its springy muscles absorb energy from the impact, which can then be used to propel it on its next jump. If we add up the energy required for each jump, ignoring this absorbed energy, the kangaroo does use more energy than it gets from its food – but by reusing energy rather than wasting it, the kangaroo achieves more than would otherwise be possible.

Electric cars use a similar principle. Instead of wasting the energy of the car's movement when the brakes are applied, squandering it by heating up its brake pads, the car uses some of the energy from braking to recharge its batteries.

38

TO STAY HEALTHY, WE NEED TO DRINK EIGHT GLASSES OF WATER A DAY

Water is good for you. This is hardly a surprise – but some of us don't drink enough fluids to stay healthy. We are often sternly told that we need to drink eight glasses of water a day (around 2 litres or 3.5 pints). And it has to be water. Forget your squash or tea. Only pure water counts – and getting through those eight glasses can be a chore. Unfortunately (or perhaps fortunately for those of us who enjoy a wider range of drinks), there is no scientific basis to any of this, other than the fact that some of us don't drink enough fluids.

Water is certainly an essential part of our diet. This reflects the significance of water to every living thing. The cells that make us up contain water, which both prevents them from collapsing and acts as a medium through which their tiny mechanisms can interact – without water, cells simply could not function. The human body contains around 60 per cent water. We can go without food for a couple of weeks, but the maximum we can survive without water is about three days. However, the necessities of how to get our hydration in an effective fashion are often distinctly misleading.

That very specific requirement to drink eight glasses a day is another myth based on a misunderstanding of a genuine scientific finding. The eight glasses figure seems to be based on a 1945 US National Research Council suggestion that our diet should include around 1 millilitre of water for each calorie of food consumed. That would add up to 2 litres a day for a typical consumption of 2,000 calories. (These days, many eat more calories still.) But this has nothing to do with drinking glasses of water.

Very little of our diet is entirely dry – this is not surprising, given the dependence of all life on water. Around half of our water intake comes in our food with no drinking required. That immediately halves the need to drink fluids. And research has shown that there is very little difference in hydration between pure water and almost any of the drinks we enjoy. Caffeine has a slight impact on how quickly we pass water, but seems to have little impact on hydration itself. Too much alcohol in a drink is more of a problem, which is why it is a good idea to avoid drinking alcohol on flights, which tends to dehydrate you anyway because of the reduced cabin pressure – but alcoholic drinks with a high water content such as beer are still effective for hydration. This is just as well, given that for centuries a weak beer known as small beer was the standard beverage in the UK, as many water supplies were unsafe to drink.

The good news if you actually like the taste of sports drinks (does anyone?) is that they too are effective at hydration, though no more so than any other drink. And despite the claims of manufacturers, it's fine to base your sports hydration on drinking when you feel thirsty. The idea of 'staying ahead of your thirst' has no scientific basis. Nor do we need our drinks to contain the electrolytes that are added to sports drinks – these chemicals are important, but we get sufficient from our food without boosting levels artificially.

While most of us are aware of the dangers of not getting enough water, even if we might have been led astray by the eight glasses guidance, less familiar is the idea that it is also dangerous to drink too much. If water is drunk to excess, body cells can swell, potentially causing brain damage and, at the extreme, death. This isn't a problem with just gulping down a glass of water, but going past a litre or so in one go starts to be risky.

39

IF YOU GET A POSITIVE RESULT FROM A 99 PER CENT ACCURATE MEDICAL TEST, THERE'S A 99 PER CENT CHANCE YOU HAVE THE CONDITION

The Covid-19 pandemic made everyone aware that the accuracy of medical tests could be evaluated numerically – but what was rarely mentioned is that those numbers alone fail to tell us what we really want to know.

Usually there are two different figures given for the accuracy of a test: its sensitivity and its specificity. The sensitivity tells us how often the test will indicate that you don't have the disease when you do (false negatives), while the specificity tells you how often the test suggests you have the disease when you don't (false positives). So the first thing to check is what's meant by saying that a test is 99 per cent accurate.

Take, for example, the lateral flow tests commonly used in the pandemic. In a large study by Oxford University, these had a high specificity – typically 90 to 100 per cent – and a lower and more variable sensitivity in the 40 to 97 per cent range. Let's say, to try the numbers, that we were dealing with a test with 99 per cent specificity and 70 per cent sensitivity.

In a false positive, the test says you have an infection when you don't, so you may have to isolate or take other precautions unnecessarily. In other tests, false positives could mean that you are given a worrying diagnosis of, say, cancer, that may cause considerable distress before it is countered, and that could even lead to unnecessary medical procedures. That '99 per cent accurate' value is here the proportion of times that the test gets things right – so in 1 per cent of cases, the test will produce false positives. Of all the positive tests, 1 per cent will be from people who aren't infected.

Unfortunately, that's not what we really want to know. The 1 per cent figure is the chance that you'll get a positive result when you aren't ill. But what we really want to know is the chance that you aren't ill if you get a positive result. This sounds very similar, but the numbers involved can be vastly different.

Imagine there are 1 million tests being taken each day, while the infection rate is, say 200 in 100,000. We need this information to apply Bayes theorem, a cunning mathematical trick for swapping round what it is that we know and what we want to know. Let's do the maths – it is surprisingly painless.

If 200 in 100,000 have the disease, then on average 2,000 of the million people tested will be infected and 998,000 won't be. The 70 per cent sensitivity tells us that 70 per cent of the 2,000 infected people will be told they have the disease – that's 1,400. And the 99 per cent specificity tells us that 1 per cent of the 998,000 people who aren't infected will get a positive result – that's 9,980 people.

In total, then, there will be 9,980 + 1,400 = 11,380 positive results. Of these, 1,400 are correct. So, the chances you have the disease after getting a positive result is 1,400/11,380 – it will be true for about 12 per cent of positive test results.

Let's say that again, because it's quite mind-boggling. Using this 99 per cent accurate test, if you are told you have the disease it will be true in 12 per cent of cases and not true in 88 per cent of cases. How accurate the test was at telling you if you have the disease after a positive test depends on how many tests are taken and how prevalent the disease is, as well as how good the test is.

Note that this does not mean we should avoid tests or ignore test results. If you take a test, it is often because your circumstances are unusual. For example, you might have symptoms, or you might have been exposed to someone who has the disease. In these circumstances, you are no longer part of the general population. The prevalence of the disease among, say, people with symptoms is far higher than it is in the population as a whole, so the starting probability is far greater than 200 in 100,000.

When routine testing is undertaken, though, the need to transfer the meaning of 'accurate' from what we can say about the test to what we can say about the disease is something that needs to be considered and that even many doctors struggle to understand. If a disease is relatively uncommon and many tests are taken, even a small inaccuracy in the test can produce a large number of false outcomes.

— 40 —

TOAST DOESN'T REALLY FALL BUTTER-SIDE DOWN

There are many humorous observational 'laws', from Murphy's Law (if it can go wrong, it will) to the Peter Principle (an individual will be promoted until they reach a position where they are incompetent). Although funny, some of these widely held rules of thumb do apply for real. For example, buses genuinely do quite often come in bunches rather than evenly spread out.

The bus-bunching effect happens because a bus that stops to pick up a lot of people will spend longer than expected at the stop. Because of this, the gap before the next bus arrives at that stop is shortened. The bus that arrives soon after may well find an empty stop and so may not need to stop at all. This continues until the front bus is too full to pick up the whole queue, at which point the gap to the third bus in line starts reducing.

However, some of these observational effects seem to be more a matter of selective memory than anything that is really happening. We are more likely to remember an event that seriously inconveniences us, or that stands out in some other way. Most of us, for example, have stronger memories of train or plane journeys where there was a serious delay than those where everything ran on time.

This is also why we are surprised by coincidences. I once bumped into someone I knew from university on a zebra crossing 200 miles away from the only place I had ever seen him before. But for the rest of my life I have passed many thousands of people while crossing roads who I didn't unexpectedly recognize. It is the case that stands out that remains in my memory. Similarly, to get a phone call from someone you have just thought about can seem amazing … until you try to work out how many times you have thought about someone and they weren't on the other end of the phone line soon after.

It seems perfectly reasonable, then, to assume that the idea that toast falls butter-side down is also a matter of selective memory. However, this is an example where the dismissal of the myth is itself a myth. Because the buttered side of toast really *is* more likely to hit the floor. This is despite the BBC doing an onscreen demonstration in which they 'proved' this was not true, showing that the toast only landed butter-side down 50 per cent of the time, just as they had expected.

If you were looking for a physical explanation of butter-down toast, you might think that it was an aerodynamic effect. Once a slice of toast has been buttered, one side of it has different properties to the other. It looks and feels different. Perhaps that might change the way that air moves over it, and hence how it falls? But the real answer doesn't require a study of aerodynamic principles.

The reason it happens – and the reason the BBC's 'experiment' was a failure – is all about the way that toast falls in the real world. The TV demonstration flipped the toast high in the air like a coin and, as with a coin, they got 50:50 results. But that's not how we end up with toast on the floor in real life. What usually happens is that the toast slips off a plate, out of your hand or off a work surface from around waist height. As one edge of the toast will usually start to fall first, it rotates. But in the time between leaving the initial, level, safe position and hitting the floor, the toast typically only has time to make half a turn. We usually start with the toast butter-side up. So, after that half-revolution, it's the buttered side that comes into contact with the kitchen tiles.

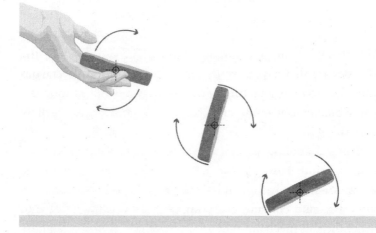

Interestingly, even a coin toss is slightly biased depending on the way up that the coin starts. Here, there is a slight probability that the side that starts face up will also end face up, as an analysis of the flight of a flipped coin shows that it spends more time in flight with the its initial top face uppermost – but the outcome, like the BBC's toast flip, is still very close to 50:50.

— 41 —

THE SUN
IS YELLOW

Whether you are a primary school child or an adult, the chances are that if you were given a pile of coloured crayons and asked to draw a picture of our friendly neighbourhood star you would colour in the sun a bright yellow. But it isn't yellow at all. It's white.

Think about how we describe an object as having a particular colour. White light falls on it and it only sends part of the spectrum back. That initial light is not yellow – it's white. The sun's spectrum contains everything from red to violet, able to make the full spectrum when raindrops split it into a rainbow. If the light were yellow in the first place, rainbows would be far less interesting.

Part of the problem we have with the sun is that it is so bright that it is dangerous to look directly at it, something we sensibly avoid. It is only when the sun is setting that we typically see its colour – and then it usually appears to be red. Yet the sun doesn't care what time of day it is. The flow of day and night has nothing to do with the sun – it is entirely dependent on the turning of the Earth. This means that any apparent change in the colour of the sun through the day must be caused by the Earth's rotation and has nothing to do with the sun itself.

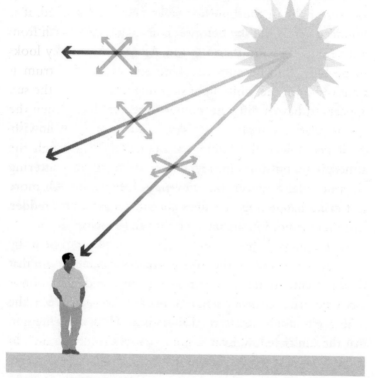

The culprit here is the Earth's atmosphere. The gases in the atmosphere are good at scattering sunlight. When this happens, a gas molecule absorbs a photon of light, then sends another photon off, often in a different direction. As it happens, the molecules in the atmosphere are best at scattering light from the blue/violet end of the spectrum and least effective at the red end. Most of the red and yellow light heads straight through the atmosphere, while a fair amount of the blue to violet energies are scattered.

We can see one obvious outcome of this scattering. Because light in the blue to violet range is scattered, it is bounced across the sky before it comes down to Earth from a different direction than the sun. As a result, the sky looks to be blue. And because this blue end of the spectrum is removed from the white light beaming down on us, the sun appears to have a different colour to its real hue. When the sun is relatively high in the sky, it does appear yellowish. As it gets lower, the light passes tangentially through the atmosphere, putting considerably more of those scattering air molecules between the viewer and the sun. With more scattering happening, the sun's appearance becomes redder. But there is no doubt that sunlight itself is white.

Astronomers, who should know better, don't help by sometimes referring to the sun as a 'yellow dwarf', a term that is almost entirely inaccurate. The sun's proper classification is a G-type main sequence star. G-types are white in colour: the yellow bit simply refers to that traditional misapprehension that the sun is yellow. Even astronomers have been known to colour it in that way. As for the star being a 'dwarf', this makes it sound as if our neighbourhood star is of low intensity and size compared to many of its peers. In reality it is relatively bright, coming in the top 10 per cent of all stars – there are a lot of dim ones out there. And its size is pretty much average.

— 42 —

THE PHASES OF THE MOON ARE CAUSED BY THE EARTH'S SHADOW

We are so used to the moon appearing in the sky at night that it can be easy to forget what a remarkable (and beautiful) thing it is. Our moon is not the largest in the solar system – Jupiter and Saturn have four bigger moons between them – but compared to the size of the Earth it is still enormous. What's more, without the moon, life may never have started on Earth.

This is because of two contributory factors. One is the way that the moon is thought to have been formed. This is likely to have been when a planet-sized object collided with the early Earth. This impact splashed out the moon and left the Earth with an unusual physical make-up, with a thin crust, enabling the escape of more of the greenhouse gases that made the Earth warm enough to live on, and an unusually large metal core, producing a strong enough magnetic field to protect us from the destructive output of the sun known as the solar wind. The other thing the moon does for us is to stabilize the tilt of the Earth, making the climate sufficiently consistent for life to form and thrive.

Our moon is remarkable in other ways too. By pure coincidence, it is about four hundred times smaller than the sun and about four hundred times closer. As a result, it pretty well exactly covers the sun at the time of a solar eclipse when the sun, the moon and the Earth are aligned, with the moon between the sun and the Earth. This coincidence won't hold for ever – the moon is gradually moving away from the Earth – but it will remain the case for some thousands of years still.

Another feature of the moon is that, apart from a slight wobble, it always presents the same face to us. The only way this can happen is if it takes the moon exactly the same time to spin on its axis as it takes to orbit the Earth. This might seem like another remarkable coincidence, but it is caused by something we usually think of as only applying to the Earth – tides. The moon's gravitational pull (with some help from the sun) causes the tides in our oceans. Because the Earth is much heavier than the moon, the moon is subject to much bigger tidal forces than we are. There is no water to pull around, but this does create a bulge in the moon's surface – which is then attracted a little more to Earth than the more distant parts. Over time, this bulging synchronized the moon's rotation speed with its orbit.

Because we only see one side of the moon, the far side of the moon is often called the dark side (Pink Floyd have a lot to answer for). It is, in fact, just as bright as the side we see. And that brings us on to those distinctive phases. When the moon is not full (all bright on the side facing us) or new (totally dark on the side facing us) it is part in shadow. It can be difficult to see, but you can sometimes faintly make out the shadowed part. It seems quite natural to assume that what we are seeing is the shadow cast by the Earth when it gets in the way of the sunlight that illuminates the moon – but it isn't.

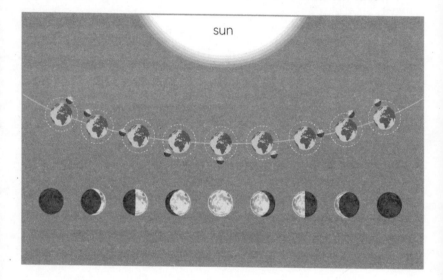

Even if the Earth wasn't here and you were floating in a spacesuit in its place (leaving aside the fact that the moon wouldn't stay in orbit), the moon would have phases. It's simply a matter of the angle at which sunlight hits it. When light hits the face that we see full on, there's a full moon. When it's hitting the opposite face full on (making the far side fully illuminated) there's a new moon. More often than not, sunlight hits at an angle and only part of the side of the moon we see is illuminated.

One last amazing moon fact – it is a lot smaller than it looks. Due to an optical illusion that still isn't fully understood, we see the moon as far bigger than it really is. This is why it looks so tiny when we take a photograph of it without using a telephoto lens. The true size of the moon as seen from Earth is about the same as the hole produced by a hole punch, held at arm's length.

– 43 –
ANTIOXIDANTS ARE GOOD AND FREE RADICALS ARE BAD

The health industry (as opposed to the more scientific field of medicine) depends heavily on loaded words, where some things are considered inevitably good and some inevitably bad. This reflects a wider tendency in the way that science is often presented to us. The media likes simple, easy descriptions of the world around us – but one of the absolutely central lessons of science is that things are never truly as simple as we first think, and we need a more detailed picture to understand what's going on.

Perhaps the most dramatic example of this complexity is in the field of climate change. To read most reporting on this, you would think that greenhouse gases are evil things that should be banned. They aren't. If the Earth's atmosphere had no greenhouse gases, our home planet would be a permanent snowball with little or no liquid water, and hence unlikely to have any life. Without the warming blanket of these gases in our atmosphere, the average temperature on the Earth would be 0 °F (-18 °C) – around 60 °F (33 °C) below current levels. We need a certain level of greenhouse gases to keep temperatures in the range that are survivable. Throughout the history of the Earth, greenhouse gas levels have varied, taking the planet from being an ice world to a steamy tropical environment. It's not that greenhouse gases are bad in themselves, but rather that we need levels to be in a certain critical range – a range that we have started to exceed since the Industrial Revolution.

Similarly, the picture with antioxidants and free radicals is one of balance, rather than a simple, cut and dried, 'antioxidants good; free radicals bad'. The first problem with this stark contrast is that while antioxidants do have an important role to play in living things, that doesn't mean that adding antioxidants to everything will make them better products. Putting antioxidants in a shampoo, for example, won't benefit your hair. And, perhaps more surprisingly, consuming a food that is 'rich in antioxidants' will not make a significant difference to the antioxidant levels in your body.

It's worth first seeing what these two types of chemical – antioxidants and free radicals – do. A radical in chemistry (as opposed to politics) is simply an atom or molecule that has one or more electrons in its outer 'valence' band, making it good at reacting with other atoms and molecules. (The 'free' bit in 'free radical' just means they aren't locked into a particular part of the body.) Free radicals do important jobs in the body, whether it is in enabling your body's defences to kill bacteria or in transmitting signals in cells. However, uncontrolled, because free radicals are so reactive, they can also do damage, potentially interfering with DNA and leading to some forms of cancer, cardiovascular issues and diabetes.

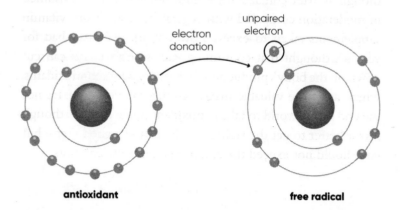

Because of this, the body makes use of antioxidants. These chemicals prevent the free radicals' oxidation processes from occurring by providing electrons to fill the gaps in the free radicals' valence band. The best-known antioxidants are vitamins A, C and E, which we get plenty of in a good diet, while the body also produces a number of antioxidants and protective enzymes to counter the unwanted effects of free radicals, such as the chemical glutathione. So, the first observation on products that are sold as being rich in antioxidants is that, with a decent diet, we don't need additional antioxidants.

There is, however, a bigger problem with the obsession with antioxidants. Pretty much everything is bad for us in excess. Even that most essential chemical for life, water, can kill if too much of it is consumed. And knocking back too many antioxidants is unhealthy too. This doesn't mean you have to cut back on, say, fruits 'rich in antioxidants' such as blueberries or cranberries. That richness is only relative. You are unlikely to cause harm by going a little over the top with these – though all fruit contains sugar and should only be consumed in moderation compared with vegetables. Antioxidant vitamin supplements taken to excess, however, are definitely bad for you. It's thought that consuming excess antioxidants can cut down on the body's production of its own essential antioxidants, which are more valuable than anything you consume. It's not the end of the world to take antioxidant supplements – though it's far better to get the right levels by eating a good diet – but you should not exceed the recommended daily amounts.

— 44 —

THE AMAZON RAINFOREST GIVES US THE OXYGEN WE NEED TO BREATHE

The world's rainforests, particularly those of the Amazon region, are poster children for saving the planet from environmental harm. These giant patches of trees are sometimes described as the planet's lungs – though strictly the image that is intended is anti-lungs. Our lungs enable us to take in oxygen and emit carbon dioxide. But the reason we like trees is that they take in carbon dioxide and emit oxygen. When this happens, it is a good thing from the point of view of mitigating global warming, as it takes carbon dioxide out of the air. Healthy rainforests are a valuable environmental asset. But it's a myth that they are necessary to produce the oxygen we breathe.

The amount of oxygen on the Earth has been roughly constant for around 2 billion years. Initially, the Earth lacked this gas – but bacteria in the early days of life pumped oxygen out into the atmosphere. For some early forms of life this highly reactive gas proved deadly, but it enabled a whole new ecosystem to develop.

It's not the case that all the oxygen would somehow disappear if the rainforest stopped producing it. Oxygen levels are slowly depleted by animals and other chemical reactions, but most of the oxygen is returned to the atmosphere from the seas. A lot of this is a result of the action of plankton that produce oxygen from carbon dioxide by photosynthesis – the same mechanism as land plants. Some of that oxygen is then reused by fish and other marine animals.

Bear in mind that the oxygen is not somehow used up by animals and new oxygen is made by plants and plankton. It is the same oxygen that was already present 2 billion years ago. The Earth is a vast recycling mechanism.

So how much does the Amazon rainforest contribute? The key here is that word in the first paragraph: 'healthy'. A good, healthy forest does help maintain the world's oxygen levels. However, the amount involved tends to be wildly exaggerated. An often-quoted figure is that the Amazon produces 20 per cent of our oxygen – and without it we won't survive. However, the Amazon is not in a good state of health. The oxygen contribution from the Amazon has never been huge, but at the moment it is around zero – and at times it is a net remover of oxygen from the atmosphere thanks to bacterial action on rotting material on the forest floor absorbing oxygen.

The 20 per cent figure is pure fantasy. Although carbon dioxide levels in the atmosphere are higher than we want them to be, this gas still only makes up about 0.4 per cent of the atmosphere, while oxygen is about 20 per cent. If the rainforest were to produce 20 per cent of our oxygen it would require the Amazon to use up far more carbon dioxide than is in the atmosphere. (And using up that CO_2 entirely would be a terrible thing. Without greenhouse gases, the Earth would be far too cold for us to live on.)

Although we love tree planting, unmanaged forests aren't the best way to produce oxygen. At night, trees take oxygen back in, so only about half of the oxygen they produce remains in the atmosphere. And, as we have seen, the rotting material that lines a forest floor contains plenty of oxygen-hungry bacteria, which absorb pretty well all of the rest of the trees' output.

The Amazon is anything but a bad thing. The rainforest is an important habitat, and it does take some carbon dioxide from the atmosphere, even though as an overall ecosystem it doesn't produce much net oxygen. But it is in no sense the world's lungs.

— 45 —

SUBLIMINAL MESSAGES IN CINEMAS WERE USED TO SELL REFRESHMENTS

An audience sits in a cinema, intently watching the movie. They don't notice that every now and then a few additional frames have been added to the film. These frames contain messages designed to get the viewers to buy snacks and drinks. And, remarkably, during the interval, sales of these sweet and salty goodies go through the roof. This was the birth of subliminal advertising: the idea that the brain would take in a message, even though it was on the screen for too short a time to be noticed. It was such an insidious form of mind control that it was banned in many countries, including the UK. Which is amusing – because the whole experiment was made up.

That word 'subliminal' dates back to the 1880s, when it was used to describe a sensory stimulus too weak to be consciously perceived. But it came to the fore in this context in 1957 when newspapers started to refer to subliminal advertising.

The entirely liminal stimulus for this sudden outbreak of news was the work of a market researcher called James Vicary, who claimed to have carried out these experiments in a US movie theatre.

Vicary wrote up his experiences in the industry magazine *Advertising Age*. He described how he had shown subliminal images of Coca-Cola and popcorn being enjoyed to over 45,000 audience members. In the interval, he claimed, sales of Coke were up 18.1 per cent and of popcorn a remarkable 57.5 per cent.

Not surprisingly, Vicary's report caused excitement among psychologists, who rushed to duplicate his findings. And failed. All the evidence is that Vicary never carried out the experiments at all and simply made the numbers up. The genuine research showed that subliminal advertising couldn't persuade people to buy something they wouldn't have bought anyway, though there was some evidence that it had a small effect.

The legitimate research suggested that if someone was already thirsty, a subliminal message could make them aware of the sensation, giving a slight impetus to act on it. And it could have a small effect in moving a thirsty cinema patron towards, say, a particular brand of cola. It's ironic that while Vicary claimed a strong increase of sales from his advertising, he said that it would not encourage anyone to switch brands.

This isn't the last word in the story, though. Research since the 2010s has shown that many studies in the softer sciences such as psychology were flawed. This has led to what's been called the 'replication crisis'. When attempts have been made to reproduce the findings of many psychology studies, only around one-third have resulted in the same findings being produced. One of the problems has been that the social sciences tend to require a far lower burden of proof than, say, physics.

A typical requirement for a result to be significant in psychology is that its 'p value' is less than or equal to 0.05. This means that if there was no real effect, the chance of observing this result (or stronger evidence) is 1 in 20. Compare this with physics, where the gold standard is a p value of 0.000003 – meaning the chance of getting such a result without a cause is about 1 in 3.5 million. Worse still, the p value suffers from the same problem as the medical test in chapter 39. It tells us the probability of getting this particular outcome if there is nothing there. We really want to know the probability of there being nothing there, given our observations. Another problem is that many of these studies were poorly designed, had too small a sample, and often had their data carefully selected to match a particular outcome.

Awareness of this problem has meant that psychology findings since about 2015 may be more reliable – but it casts doubt on findings from when subliminal advertising was widely studied. It's arguable that whether or not it works it was right to ban it – because it was an attempt to coerce people without their knowledge. But the chances are that it had little or no impact.

— 46 —

COCKROACHES CAN SURVIVE A NUCLEAR BLAST

Viewers of sci-fi movies in the 1950s had a pretty good picture of the outcome of nuclear blasts. Most living things would be wiped out, while some would survive, mutating into monsters. These might be giant ants or killer rodents. Perhaps even the odd radioactive spider, ready to create your friendly neighbourhood Spider-Man. But one thing everyone knew for sure was that cockroaches would be among the survivors. This is even referenced in the Pixar movie *Wall-E*, where the titular post-apocalyptic robot has a pet cockroach called Hal. (This seems to be a dual in-joke, referring to both the HAL-9000 computer in the movie *2001: A Space Odyssey* and the Laurel and Hardy producer Hal Roach.)

One way of checking out this picture for real is in the countryside around Chernobyl in Ukraine. The Chernobyl nuclear power station disaster was the result of a combination of human error and bad systems. On 25 April 1986, while testing an emergency cooling system, an operator nearly shut down the reactor. Wanting to keep the energy flowing, engineers overrode the safety system and pushed up the reaction rate. As the temperature suddenly shot up, the steam pressure became too great and blew the reaction vessel open.

Radiation spread across a wide area, which was abandoned by humans. But rather than hosting monstrosities, the forests around Chernobyl are full of healthy wildlife. It's true that many animals died – but those that did survive went on to breed normally. Almost certainly, those surviving animals would have included cockroaches. But not because of any ability to survive a nuclear blast. (In practice, the Chernobyl accident was not strictly a nuclear blast, it was the result of steam pressure, releasing radioactive materials.)

Cockroaches are certainly survivors. These common insects, related to termites and mantises, are found in many different habitats with over four thousand different species known, but of these only a few dozen tend to co-habit with humans – pests that live off our dropped food, particularly prevalent in commercial kitchens.

It is the case that the different species of cockroach include those that can cope with pretty well every climate on Earth, from the polar regions to the tropics. They have also been around far longer than we have. Humans have been in existence for around three hundred thousand years, with other hominids going back a few million years. But the predecessors of cockroaches date back well before the near-total extinction of the dinosaurs, stretching back over 300 million years.

There is no doubt also that cockroaches are robust insects, but they don't hold the record for surviving unlikely challenges.

This goes to the tardigrade or water bear – strange little creatures around half a millimetre long that are distantly related to centipedes. Tardigrades have survived extremes of heat and cold, dehydration, lack of food, and exposure to outer space. Some cockroaches are good in many of these situations – but to a far lesser extent. For example, cockroaches have survived for up to twelve hours at around -6 °C (21 °F). Tardigrades have lasted thirty years at -20°C (-4 °F) and remarkably for days at -200°C (-328 °F).

The idea that cockroaches would outlive us in the event of a nuclear blast seems to date back to the aftermath of the nuclear attacks on Japan in 1945, when cockroaches were seen among the rubble. While it is true that cockroaches can survive more radiation than we can – it would take around ten times as much radiation to kill them – their abilities are only moderate compared with those of the tardigrade. These remarkable creatures can stand up to a thousand times our limit. So, if any organism is going to take over should we blast ourselves into extinction, it's far more likely to be the water bears.

— 47 —
FISH IS BRAIN FOOD

Fish is a healthy alternative to meat, and since time immemorial, parents have tried to persuade children to eat what is often an acquired taste by claiming it is 'brain food'. Exactly why this used to be the case is not entirely clear, but more recently the claim has been linked to the presence of Omega 3 oils in both oily fish and fish oil supplements.

Omega 3 oils are those where there is a particular double bond between two of the carbon atoms in the molecule, occurring three places from the end of the molecule (hence the 3). These oils are prevalent in fish oil, but also turn up in some nuts and seeds. It is generally considered far better to get these oils by eating oily fish than supplements. (Apart from anything, fish oil supplements tend to have a considerable amount of vitamin A, which is easy to overdose on.)

One of the problems with making a good assessment of claims of benefits on mental ability from diet is that it is very difficult to separate diet from other factors. For example, a family that has a good diet may also provide all sorts of other things for their children that support mental development. Being sure that the diet is the cause is difficult and many of the studies have been poor – with small numbers involved, or not controlling properly for other factors. This has not stopped the media repeatedly claiming that there is good evidence that fish, and particularly Omega 3 oil, helps the brain.

In the UK, two trials have had considerable coverage in the press, both based in the north-east of England. One was relatively small with around a hundred children, using Omega 3 oil capsules, but was quite well run in terms of making sure that there was proper 'double blinding' to ensure both children and researchers could not be influenced by expectations. The BBC reported on this before the results were published and claimed a dramatic improvement in one student named Elliot who had been falling behind. 'Over the past year, a dramatic change has taken place in Elliot,' the BBC gushed. 'He has soared through the Harry Potter books and now heads to the library after the school bell has sounded.' This was irresponsible reporting – given the trial was double blinded, it's entirely possible Elliot wasn't even taking Omega 3. And when the results were published, they did not show a useful benefit overall.

The second trial seems much better in scale. It involved 3,000 children and did support the benefits of fish oil. However, the trial was run by a fish oil company and was not controlled (so did not compare those who took the oil with those who didn't). Over two thousand of the children dropped out before the trial was finished. Perhaps worst of all, the researchers involved went into the exercise expecting to find positive results – and the data published only covered some of those taking part. This is known as cherry-picking and totally invalidates a trial.

The world's leading body for assessing clinical trials, the Cochrane Reviews, has failed to pick up any significant benefit from taking fish oil supplements. There even seemed to be small *negative* impacts for older people and for the babies of pregnant women who took the supplement.

So, fish really isn't brain food. Only two dietary items seem to have measurable benefits on cognition. One is a small improvement in babies from consuming breast milk. The other is that coffee (not caffeine alone, but specifically coffee) appears to improve mental capacity a little in the middle-aged.

— 48 —

TV AND FILMS SHOW MOVING PICTURES DUE TO PERSISTENCE OF VISION

The kind of moving picture we watch in the cinema, on the TV and in internet videos is odd, when you think about it. The pictures don't really move at all. What is put on the screen is a series of still images or 'frames'. Typically, these are shown at between 24 and 50 times a second. And yet we perceive something that moves like the real world.

The technology to do this dates back to before photography. Toys with entertaining names such as the thaumatrope, the phenakistiscope and the zoetrope became available in the 1820s and 1830s. These made use of spinning discs or cylinders with step-by-step images, usually drawn, which are glimpsed one at a time in fast sequence, producing the impression of motion. When the British-American photographer Eadweard Muybridge produced sequences of photographs taken in quick succession of moving animals and people, these were first viewed in a zoetrope and then projected by Muybridge using a device he called a zoopraxiscope from 1879.

The explanation for how this trickery worked, which emerged around the same time as projected moving pictures and would remain generally accepted as an explanation right through the twentieth century, was 'persistence of vision'. The idea was that each rapidly displayed still image lingered in the brain for a sufficient time to bridge the gap until the next image was shown and the two images were merged together, making it look as if the difference between the two was due to motion.

However, this explanation has now been totally discarded. There are two problems with it. We do get a visual after-image, but this appears not to form until around 50 milliseconds after the image disappears – and that isn't quick enough to bridge the gap between images and avoid a visual flicker. For that matter, the whole idea of persistence makes no logical sense as a way to produce apparent motion. If you overlay two or more images from a cinema film you don't get motion, you get a blurry mess.

Instead, the appearance of moving pictures is just one of the many optical illusions produced by the tricks that the brain and eye combination use to cover up potential problems with our vision. We don't see the world around us in the same way that, say, a video camera detects a scene. Our brains make note of various aspects of the incoming signals passing from the optic nerves in the eye, picking out details such as shapes, edges, colour and shading contrasts. From this data, we assemble made-up pictures of what the world looks like.

When you consider what the eye and brain deal with, this isn't surprising. For example, there is a patch on the back of the eye with no receptors – we should have a blind spot, but the brain uses information from both eyes to recreate the missing part. And for that matter, our eyes dance around in very rapid movements called saccades. If we were to see what the eyes see, the view would constantly be jiggling around, far worse than the most amateurish hand-held video. Our brains edit all this out, creating an apparently steady but not strictly real viewpoint.

The reason we see a sequence of still pictures as smoothly moving on the screen is nothing to do with persistence of vision. It is an accidental side effect of the way that the brain's visual modules assemble elements to create a consistent overall picture. Most optical illusions are a momentary bit of fun – or just confusing, like the illusion that makes the moon look bigger than it really is. But the moving picture optical illusion makes it possible to produce one of our favourite forms of entertainment.

– 49 –

EVOLUTIONARY CHANGES TAKE MILLIONS OF YEARS

The reality of evolution remains divisive among some parts of society – but scientifically speaking there is no controversy. Evolution is how species develop. It is simple common sense. Living things pass on characteristics to their young, and these characteristics vary from individual to individual – we now know because of genetic variation. How well an individual will survive in a particular environment depends on these characteristics – so it's hardly surprising that those with characteristics that help them survive are more likely to pass them on to their offspring.

Many of the opponents of evolution accept the idea of 'microevolution' – that a species can vary to survive better in a particular environment – but aren't happy with the idea of new species evolving from old ones. This reflects an oddity of the way that species develop and change. Scientifically, each offspring is the same species as its parents. So, you might expect that it is indeed impossible for a new species to evolve. But in practice this is not a problem.

A useful analogy is the colours of the rainbow. As we've seen, there are far more than seven of these, with over 16 million on a typical computer. If you look at two of the adjacent colours in the 16 million plus, it is impossible to distinguish them by eye. A colour is always the 'same' colour as its neighbour – yet travel across the whole spectrum and colours clearly change drastically. Like the colours, a species isn't a set, definitive thing, but rather an accumulation of characteristics. When there is enough change in these characteristics you get a new species.

Major evolutionary changes – changing, for example, from bacteria-like organisms to a mammal like us – do require an extremely long timescale. Darwin was aware of the need for this kind of time period and early on in the development of evolutionary theory he was running ahead of the physics and geology needed to support his ideas. For example, until it was understood how the sun works by nuclear fusion, it was assumed that it was literally burning, something that would not be possible for more than a few million years – insufficient time for the panoply of evolution that we see to emerge. We now know that the Earth has been around for 4.5 billion years, and life has existed for much of that timescale.

It is easy to mentally limit evolution to these 'evolutionary' timescales. But evolution is happening all the time and can have visible results in surprisingly short periods of time. A great example of this is a process known as 'industrial melanism' typified in an insect called the peppered moth. Common in Europe, the moth has evolved a colouring that enables it to blend in well with the mottled surface of lichen-covered trees. Moths that are hard to spot are less likely to be picked off by birds and eaten.

During the Industrial Revolution, the tree bark favoured by the moths became darker in industrial regions. This was partly down to soot, and partly because pollutants killed off the lichen, exposing the darker bark beneath. Within a few generations the moths were getting darker – not just a little, but drastically so. There was always some variation in the colouration; with the blackened bark, moths with darker colouring were more likely to survive predation and be able to breed.

Since the introduction of clean air legislation, buildings in industrial towns have become cleaner and brighter – and so have the trees. As a result, the moths have reverted to their old colouration. This was evolution in action over decades, not millions of years. Similar outcomes have been observed with birds known as Darwin's finches in the Galapagos Islands. Depending on weather and climate changes, different types of plants do well here. When the most common seeds have been large and hard, birds with big beaks have thrived. When heavy rainfall has made small seeds more common, an ever-increasing number of birds have smaller beaks. This process has been observed over periods of time of less than a decade. Evolution is always with us, always happening.

— 50 —

SCIENCE WORKS BY PROVING THEORIES THAT ARE TRUE

It's common to think of scientists as nature detectives who come up with deductions about the deeper realities of the world from the observations they make, enabling them to prove what had been mere theories up to that point. This picture is both incorrect and potentially dangerous for a society that is dependent on science for its survival.

Science works by finding patterns. If the universe behaved differently every time something happened there would be chaos with no underlying patterns and there could be no science. We can, of course establish some simple scientific truths. If I say that there are five marbles in a box, it is easy to establish whether or not my theory is true. But in itself, merely labelling something doesn't advance science. We need to be able to establish those underlying patterns – in this case, *why* there are five marbles in the box.

It is, however, very difficult to prove a pattern holds in all circumstances. If I have a theory that, say, objects roll downhill under gravity rather than uphill, I can't check every single object in the universe to be able to deduce that this is true. What scientists do here, rather than making a deduction that establishes a fact, is to employ *induction*. This makes use of my observation that every time I've seen something rolling on a hill influenced only by gravity it has always rolled *down*. If the object appears to roll *up* I need to check for an optical illusion. Using induction, I can only discover whether a theory continues to be supported, or whether it has a problem, but I can never establish the absolute truth.

A famous example of the limits of induction is the theory that swans are white birds. There was a time when everyone in Europe who had ever seen an adult swan could confirm that they were white. This made it perfectly acceptable to establish by induction the theory that all swans are white ... until someone travelled to Australia and saw their first black swan. At this point, the theory had to be revised.

With anything more complicated than labelling a thing, science relies on discovering those underlying patterns. A scientist will come up with a theory or model that explains a particular observed pattern and predicts the possibility of future patterns. A theory is a description of why something is happening, whereas a model is a simplified mechanism that produces similar outcomes to a real world phenomenon that we don't fully understand. Historically, models were usually mechanical, whereas now they tend to be mathematical.

If the predictions of the theory or model come true, then it is upheld, but if it fails, it needs to be replaced or enhanced. Of course, scientists are human, and can stick with theories or models that they have spent their careers working on even when the consensus has moved away from them. But, overall and with time, the scientific method is very successful at coping with new evidence that contradicts existing theories.

This means that (again, for anything more complex than labelling things) science is not in the business of establishing the truth, because that is never going to be fully possible. Instead, what scientists do is to establish the best theories and models of what is happening *given the current data*. We always have to be ready to change our view if new data comes to light that contradicts the theory.

A particular confusion arises from the use of the word 'theory'. So, for example, the theory of evolution is often dismissed by those who feel that it challenges their views on a creator as 'just a theory'. This is because, in common usage, a theory is something weak and as yet unsubstantiated. But scientific theories that are widely supported are the best explanations we have as yet – not 'just a theory' but, rather, 'proudly a theory'.

Science has done a huge amount for us. It has given us enormous advances in everything from medicine to the quantum physics required to produce modern electronics. And it has provided fresh ideas and an ability to grasp the big picture, from cosmology's suggestions of how the universe began to the pursuit of the Higgs boson and the understanding that the theory behind it gives us of particles and their mass. But science can never be about proving that theories are true facts. It will always evolve.

INDEX